Systems Architecting

Systems Architecting
Methods and Examples

Authored by Howard Eisner

CRC Press
Taylor & Francis Group
Boca Raton London New York

CRC Press is an imprint of the
Taylor & Francis Group, an **informa** business

CRC Press
Taylor & Francis Group
6000 Broken Sound Parkway NW, Suite 300
Boca Raton, FL 33487–2742

CRC Press is an imprint of Taylor & Francis Group, an Informa business

ISBN-13: 978-0-367-34766-6 (hbk)
ISBN-13: 978-0-367-34592-1 (pbk)

Library of Congress Cataloging-in-Publication Data
Names: Eisner, Howard, 1935– author.
Title: Systems architecting : methods and examples / by Howard Eisner.
Description: Boca Raton, FL : CRC Press/Taylor & Francis Group, 2019. | Includes bibliographical references and index. | Summary: "This book provides a new approach to systems architecting, not previously available. The book provides a compact innovative procedure for architecting any type of system. "Systems Architecting: Methods and Examples" describes a method of system architecting that is believed to be a substantial improvement over "methods" previously covered in other systems architecting books. With the book's relatively straightforward approach, it shows how to architect systems in a way that both developers and clients/customers can readily understand. It uses one of the essential principles suggested by Rechtin and Maier, namely, Simplify, Simplify, Simplify. Systems engineers, as well as students taking systems engineering courses will find this book of interest"– Provided by publisher.
Identifiers: LCCN 2019023659 (print) | LCCN 2019023660 (ebook) | ISBN 9780367345921 (paperback) | ISBN 9780367347666 (hardback) | ISBN 9780429327810 (ebook)
Subjects: LCSH: Systems engineering. | System design.
Classification: LCC TA168 .E3873 2019 (print) | LCC TA168 (ebook) | DDC 620/.0042--dc23
LC record available at https://lccn.loc.gov/2019023659
LC ebook record available at https://lccn.loc.gov/2019023660

Visit the Taylor & Francis website at
www.taylorandfrancis.com

and the CRC Press website at
www.crcpress.com

Dedication

This book is dedicated, first, to Eberhardt Rechtin, who pioneered the investigation of architecting systems with his book on "Systems Architecting" and his unique way of exploring and thinking about architecting. Second, a dedication is deserved by Andy Sage, who contributed mightily to the overall field of systems engineering, which contains as a subset the matter of systems architecting.

Moving to a more personal level, I dedicate the book to my hundreds (possibly thousands) of Master's and Doctoral students during my 24 years of teaching at The George Washington University. A very large percentage of them were subject to the architecting methods devised by the author and described herein, and appeared to happily follow the suggestions and procedures set forth.

Even more personal dedications are offered to my wife, June Linowitz, who knows how to "architect art," in very creative ways. Further dedications are suggested for my children Oren David Eisner and Susan Rachel Eisner Lee, and their children Ben, Gabriel, Jacob, Rebecca, and Zachary.

Contents

Foreword

Systems architecting is a quite important part of a set of activities known as systems engineering. The International Council on Systems Engineering (INCOSE) has several definitions of systems engineering, one of which is [1]: "Systems engineering is an iterative process of top-down synthesis, development, and operation of a real-world system that satisfies, in a near optimal manner, the full range of requirements for the system". The set of elements of systems engineering might well be considered from the list below [2]:

1. Needs/Goals/Objectives
2. Mission Engineering
3. Requirements Analysis/ Allocation
4. Functional Analysis/ Decomposition
5. Architecture Design/ Synthesis
6. Alternatives Analysis/ Evaluation
7. Technical Performance Measurement
8. Life-Cycle Costing
9. Risk Analysis
10. Concurrent Engineering
11. Specification Development
12. Hardware/Software/ Human Development
13. Interface Control
14. Computer Tool Evaluation and Utilization
15. Technical Data Management and Documentation
16. Integrated Logistics Support
17. Reliability, Maintainability and Availability (RMA)
18. Integration
19. Verification and Validation
20. Test and Evaluation
21. Quality Assurance and Management
22. Configuration Management
23. Specialty Engineering
24. Preplanned Product Improvement
25. Training
26. Production and Deployment
27. Operations and Maintenance
28. Operations Evaluation
29. System Disposal
30. Systems Engineering Management

Although this is a long and formidable list, the architecture design and synthesis stands out as particularly critical since it establishes the basic structure of the system. It is that element that this book addresses.

References

1. "Systems Engineering Handbook," INCOSE, Version 3.2.1, January 2011.
2. Eisner, H., *Essentials of Project and Systems Engineering Management*, 3rd Edition, John Wiley, 2008.

Preface

A System's Basic Structure is its Architecture, and its Architecture is its Essential Framework

This book might be considered a follow-on to two breakthrough books and a massive investment by the government. The two books are Rechtin's *Systems Architecting*, published by Prentice-Hall in 1991, and Rechtin and Maier's *The Art of Systems Architecting*, published by the CRC Press in 1997. The massive investment was undertaken by the Department of Defense (DOD) when they discovered the need for a system architecture and set out to define an architectural framework, starting with the C3ISR (Command, Control, Communications, Intelligence, Surveillance, and Reconnaissance) domain and community. These books and the DoD investment have proven to be extremely valuable, and very important in terms of the overall field of system architecting. As suggested, they represent points of departure for this treatise, which moves forward, but perhaps in a slightly different direction.

There are huge numbers of information technology (IT) systems deployed in the industrial, defense, and academic sectors, each of which needs to be designed, developed, and operated. A key element of the design phase is called preliminary design or system architecting. Each and every system needs an architecture, placing this system element in a critical position. These systems also can be recognized by their functional decomposition, which has the following components:

1. Input
2. Output
3. Processing
4. Storage, and
5. Security.

Each of these systems has various sub-components that need to be chosen for each and every implementation.

Back in 1995, the DoD formally recognized that it needed a common approach to architecting these systems within its purview. This led to the so-called DoD Architectural Framework (DoDAF) approach, which has been strong and constant since then and over the years. This was a wise step despite the fact (according to this author) that the DoDAF has a few wrinkles that need to be noted and accounted for. This book presents a method of architecting that can be considered complementary to the DoDAF approach and also one that provides an alternative basis in logic.

The four-step architecting process defined here is considered to be another move forward in the complex domain of systems architecting. The author hopes that it will be used and tested by the community in the years to come.

Howard Eisner
Bethesda, Maryland

Author Biography

Howard Eisner spent 30 years in industry and 24 years in academia. In the former, he was a working engineer, manager, executive (at ORI, Inc. and the Atlantic Research Corporation), and president of two high-tech companies (Intercon Systems and the Atlantic Research Services Company). In academia, he was a professor of engineering management and a distinguished research professor in the engineering school at The George Washington University (GWU). At GWU, he taught courses in systems engineering, technical enterprises, project management, modulation and noise, and information theory.

He has written nine books that relate to engineering, systems, and management. He has also given lectures, tutorials, and presentations to professional societies (such as INCOSE – International Council on Systems Engineering), government agencies, and the Osher Lifelong Learning Institute (OLLI).

In 1994, he was given the outstanding achievement award from the GWU Engineering Alumni.

Dr. Eisner is a life fellow of the IEEE (Institute of Electrical and Electronics Engineers) and a fellow of INCOSE and the New York Academy of Sciences. He is a member of Tau Beta Pi, Eta Kappa Nu, Sigma Xi, and Omega Rho, various honor/research societies. He received a Bachelor's degree (BEE) from the City College of New York (1957), an MS degree in electrical engineering from Columbia University (1958), and a Doctor of Science (DSc) degree from the GWU (1966).

Since 2013, he has served as professor emeritus of engineering management and a distinguished research professor at the GWU. As such, he has continued to explore advanced topics in engineering, systems, and management.

Author Biography

Other Books by the Author

Computer-Aided Systems Engineering

Reengineering Yourself and Your Company

Managing Complex Systems – Thinking Outside the Box

Essentials of Project and Systems Engineering Management

Systems Engineering – Building Successful Systems

Topics in Systems

Thinking – A Guide to Systems Engineering Problem-Solving

chapter one

Background

The basic structure of a system is, by definition, its architecture. In this book, we describe a procedure for architecting a system. This procedure can be applied rapidly, is definitive, unambiguous, and critical to the process of preliminary system design.

For purposes of this treatise, there are two types of architects and two corresponding types of architecting processes. The first is the relatively widespread field pertaining to the architecting of buildings of various types. In this field we will see the prominent names and creations of the likes of Frank Lloyd Wright, Frank Gehry, I. M. Pei, and Mies van der Rohe. This type of architecting is distinctly not what this book is about.

The subject of this book is the other type of architecting, that which pertains to building *systems*. A short list of types of these systems includes:

- Defense systems
- Health systems
- Information systems
- Transportation systems
- Security systems
- Communication systems
- Human resource systems, and
- Space systems.

The body of knowledge that relates to the design, construction, and operation of such systems is generally recognized as *systems engineering*. This field has some 30 elements [1 – Chapter 7], one of which is system architecting. Some practitioners equate system architecting with preliminary design, and we accept this notion as the focus of this book which answers the question – what is the specific and recommended process by which one architects a system?

The Department of Defense (DoD)

Since the DoD is "in the business" of building many of the types of systems cited above, it makes sense to take a look at what the DoD has done by way of developing methods and procedures whereby one architects these systems. Surely there is a single system architecting approach, established by the DoD, that would apply to all types of systems.

Back in the 1990s, the DoD started to work on this issue, within the C3ISR (command, control, communication, intelligence, surveillance and reconnaissance) community [2]. The evolved approach became the DoDAF, the DoD Architectural Framework. The centerpiece of this approach was the notion that there are three critical *views* of systems that must be considered, i.e.,

- The *operational* view
- The *systems* view, and
- The *technical* view.

These views were defined (approximately) as:

- The operational view addressed how the system would perform in an operational setting and environment.
- The systems view considered systems and interconnections to support warfighting functions.
- The technical view dealt with sets of rules with respect to the arrangement and interactions between system elements.

As the structure of DoDAF expanded with time, the DoD basically required those working on systems architecting to consider many more essential views such as those articulated below.[1]

AV – 1: Overview and Summary Information
AV – 2: Integrated Dictionary
OV – 1: High-Level Operational Graphic Concept
OV – 2: Operational Node Connectivity Description
OV – 3: Operational Information Exchange Matrix
SV – 1: System Interface Description
TV – 1: Technical Architecture Profile

This made a lot of sense, in principle, and was in consonance with our tendency to "drill down," once we have found a top-level structure that we find satisfactory to our needs. Since the three basic views remained mostly unchallenged in their contribution to an architectural framework, the more one defines and drills down from there, the better. Or so it seems.

There are several unanswered questions and areas of concern with respect to the DoDAF approach. These are addressed mostly in Chapter 9 which is devoted to a more-in-depth consideration of DoDAF. For now, we will consider it a very important building block in the system architecting process, as formulated by the DoD.

Eberhardt Rechtin

A prominent position in the field of system architecting was established by the master engineer, Eberhardt Rechtin. Indeed, he wrote what this author would consider to be the seminal work in the field [3]. Here are some of the important points he made in his book:

- There are basically four approaches to the process of architecting, namely normative, rational, argumentative, and heuristic.
- Heuristics are very important in terms of designing and building new systems. One of the most important is the KISS (keep it simple, stupid) approach, and Rechtin includes a whole Appendix on his list of heuristics.
- "[T]he greatest architectures are the product of a single mind."
- It is important, also, to focus especially on boundaries and interfaces.
- It's also necessary to place extreme requirements under constant challenge.
- "The essence of architecting is structuring, simplification, compromise and balance."
- Prototyping plays an important role in designing and building the best possible system.

Beyond Rechtin's work cited above, he later teamed with Mark Maier [4] to continue ground-breaking explorations of the field of architecting and related matters. Here are three noteworthy quotes from that book:

> "architecting is creating and building structures"
>
> "the foundations of systems architecting are a systems approach, a purpose orientation, a modeling methodology, ultraquality, certification, and insight"
>
> "one insight is worth a thousand analyses"

Author Information

Going back into the 1970s, this author worked on several large-scale systems in the fields of defense, space, and transportation. Preliminary design, equated here approximately with system architecting, was a topic of great interest. Indeed, moving some 40 years down the road to today's world, interest has increased as we try to build cost-effective systems within changing, and often confusing, system acquisition environments and procedures.

The essence of a recommended system architecting process has evolved over the years and has been described, in some detail, in this author's book in 2008 [1]. This process has been tested hundreds of times through the efforts of this author's graduate students in systems engineering. Later chapters in this book provide further rationale and additional detail that is appropriate to a definitive system architecting process.

Three Special Experiences

Nimbus

One of the systems worked on by the author, going back to the 1960s, was the Nimbus meteorological satellite. This was a follow-on to the TIROS system, and was managed by the Goddard Space Flight Center of NASA. Unlike TIROS, which was a "spinner," Nimbus was a three-axis stabilized system that "looked" down at the Earth from about 450 nautical miles in space. Although there were many lessons learned from participating in this program, it was first noted that Nimbus had a variety of subsystems, such as stabilization and control, structure, thermal control, communications and data handling, and various measurement instruments that constituted the payload. It was noted also that these subsystems approximately corresponded to the "functions" to be carried out by the satellite system.

Mallard

Also in the 1960s, this author had a role on a communications system known as Mallard. This was a tactical communications system that was managed by an Army facility at Fort Monmouth, New Jersey. Mallard was a quite sophisticated battlefield communications system and this author was a sub-contractor on the GT & E/IBM team. GT&E was the prime contractor and took the lead in all systems engineering and architecting activities. Special attention was paid to the architecting approach. First and foremost, the architecting team set forth a detailed "functional decomposition" of the Mallard system. They then proceeded to design each of the key elements of Mallard with respect to each of the decomposed functions. This very successful and sensible approach was not lost on this author. It was clearly the right way to go.

Aviation Advisory Commission Design

Yet another important project for this author was serving as a consultant to the Aviation Advisory Commission as they looked at the future of our National Aviation System (NAS). The Commissioners and their Executive Team designed alternative future systems, based upon a preliminary set of functions for such systems. They then configured specific subsystems and carried out an overall evaluation of these subsystems. It was a "modified and tailored" systems engineering approach and was groundbreaking in its scope and method. These were the names of the alternatives that were configured and evaluated [5]:

1. Extension of Current Operation
2. High-Density Short-Haul Supplement
3. Remote Transfer Airport Supplement
4. Local Terminal and Exchange-port Supplement.

Here again, the approach was facilitated by a careful definition of functions and how to instantiate these functions. This "functional" decomposition idea was to become a cornerstone of how to do the part of systems engineering known as system architecting.

A Bottom Line

With the special importance of functional analysis established as a key element of systems architecting, we now set forth the critical top-level steps of this process, as follows:

1. Functional Decomposition
2. Design Approaches to Instantiate All Functions and Subfunctions (Synthesis)
3. Evaluation of Alternatives (Analysis)
4. Selection of Preferred Alternative (Cost-Effectiveness Assessment).

These then become the basic steps in architecting a system, and are discussed and illustrated in considerable detail in the remainder of this book.

Note

1. Excluding 18 Additional Supporting Views.

References

1. Eisner, H., *Essentials of Project and Systems Engineering Management*, 3rd Edition, John Wiley, 2008.
2. C4ISR Architecture Framework, version 2.0 (1997), Washington, DC, DoD, December 18.
3. Rechtin, E., *Systems Architecting – Creating and Building Complex Systems*, Prentice-Hall, 1991.
4. Rechtin, E. and Mark Maier, *The Art of Systems Architecting*, CRC Press, 1997.
5. Eisner, H., *Computer-Aided Systems Engineering*, Prentice-Hall, 1988, p. 241.

[illegible] such systems. They then configured specific subsystems and [illegible] an overall [illegible] of these subsystems. It was [illegible] and [illegible] "systems engineering" approach and was [illegible] [illegible] method. [illegible] [illegible] were [illegible] and evaluated [5]:

1. [illegible] of Current Operation
2. High-Density [illegible] Supplement
3. Remote Transfer for Airport Supplement
4. [illegible] Terminal and Exchange [illegible]

[illegible] again, the approach was facilitated by [illegible] and how [illegible] that these [illegible] "look" [illegible] [illegible] their [illegible]

[illegible]

[illegible] of systems [illegible] new [illegible] the [illegible] steps of this process, as follows:

1. [illegible]
2. [illegible]
3. [illegible]
4. Selection [illegible]

[illegible]

Note

1. [illegible]

References

1. [illegible]
2. [illegible] version 2.0 [illegible] Washington, D.C. [illegible]
3. [illegible] Building [illegible] [illegible] 196[illegible]
4. [illegible] CPC [illegible] 1997.
5. [illegible] [illegible] 1968, p.

chapter two

Purpose and Features

The systems architecting process, described in later chapters, is intended to be a fundamental skill area of the systems engineering team. Put another way, systems architecting is a critical part of the overall systems engineering activity of the enterprise. It is also seen, in this book, as more-or-less the same as preliminary design for the system in question, although some may argue that the two are quite different.

Purpose

As the systems engineering team begins the design and development of a new system, an early activity is to formulate an architecture for that system. Thus, one essential purpose of system architecting is to come up with the *preliminary design* for that system. That design is broad and inclusive, setting the stage for a deeper and more detailed process of synthesis for the system.

In many cases, the design team is in a company that is competing for government contracts. In that context, it is very important to be able to architect a system as part of the proposal process. This often means that it has to be carried out within a 30–60 day time period, and that it be transparent, clear, technically compelling, and highly competitive. This makes it possibly the most important part of the proposal process. It is directly connected to winning a higher percentage of proposals which is recognized as an important goal for the overall enterprise. Indeed, it may be the difference between success and failure for that enterprise.

Features

Some of the desired features of the architecting process can be described as:

1. Technically compelling and appealing
2. Includes a choice among alternatives
3. Able to be carried out within 30–60 days by existing personnel
4. Built upon the system's functional decomposition
5. Based upon an unambiguous and definitive process
6. At an appropriate level of detail
7. Consistent with the "systems approach."

The *technical content*, when looked at by several technical personnel, should "ring true" and be appealing. These personnel should find it simple and leaning toward elegance. We will also insist that the process explicitly include the *definition of several alternatives*. Ultimately, the preferred architecture will be selected from among these alternatives. In the context of proposal writing, the entire architecture needs to be *formulated within a 30–60 day time period*. In some cases, this time period may need to be compressed. As suggested earlier in this treatise, there are compelling reasons for the process to start with the *functional decomposition* of the system. One needs to be careful to decompose at the appropriate level of detail. Often, too many levels lead to poor results and can represent a fatal error. The architecting process must be *specific enough* so that its products represent the architecture itself. This means that the steps of the process are unambiguous. The architecting procedure, as suggested above, should not have too many levels of decomposition. *Two to three levels* (at most) are recommended. Finally, the overall process needs to follow the "systems approach." The meaning of this term is discussed below.

The Systems Approach

An excellent point of departure for examining the systems approach is the definition provided by NASA [1], namely:

> the systems approach is the application of a systematic, disciplined engineering approach that is quantifiable, recursive, iterative and repeatable for the development, operation and maintenance of systems integrated into a whole throughout the life cycle of a project or program.

There are some 12 aspects to the systems approach, each of which is defined and explored briefly in the following text [2].

1. *Systematic and Repeatable Process*. The process(es) employed in building systems needs to be systematic and repeatable, in distinction to haphazard and invented on the spot. Even though we have many brilliant engineers and scientists working on our systems, they must still fit into a disciplined environment in order to make the overall system and work force operate as an efficient team.
2. *Interoperability and Harmonious Operation*. The various parts of the system (elements, components, subsystems) must interoperate and exhibit harmonious behavior. They must be designed to do so whether or not they are being built from scratch or are considered "off-the-shelf."

3. *Explicit Consideration of Alternatives.* In effect, and especially in terms of architecture, we must design several alternatives and pick the best one from among these alternatives. This will remain a principle of design at the top-level (architecting) and lower levels (subsystem design).
4. *Iterations to Converge and Refine.* For large-scale systems, we recognize that information about the system comes to us over time, as per the results of tests as well as the search for new design information. We accept what we have, and use a TBD (to be determined) to be a place holder for new and needed information. As this information becomes available, we continue to update, refine, and converge.
5. *Robust and Slow-Die System.* Our system must not be susceptible to single-point failures, wherever possible. When failures do occur, the system must continue to operate, but in a series of degraded modes. This generally means that we must use redundancy and back-ups in order to achieve this mode of behavior.
6. *Satisfaction of Requirements.* This refers to the final set of agreed-upon requirements for the system. We accept the notion that many requirements might turn out to be negotiable during the design process. Ultimately, though, we settle in on a set of true requirements that both the sponsor and developer have accepted.
7. *A Cost-Effective Solution.* This is intuitively clear and also implies that we must be able to measure both costs and effectiveness, even at the architecting level. These measurements become better and better as we move through the development cycle.
8. *A Sustainable System.* In today's world, we recognize that systems, in general, need to be designed to be sustainable over their life cycles as well as extended lifetimes. This is another fundamental principle of design.
9. *Appropriate Technology and Risk.* Many of our more complex systems use advanced technologies (e.g., sensors, processors) in order to satisfy the system requirements. However, as these technologies are employed, the risk for the system tends to increase. Each system must be designed at an appropriate point in the technology–risk tradeoff.
10. *Architect for System Integration.* We recognize, in advance, when and where the elements of the systems will ultimately need to be integrated. We facilitate this part of the process by anticipating the need.
11. *Consider Points of View of All Stakeholders.* There are usually many different stakeholders for large-scale systems. Part of the systems approach is to understand that these stakeholders are literally "part of the system." By such broad consideration, we try to come to a better overall solution.

12. *Use of Systems Thinking.* We have learned the meaning of "systems thinking" as well as how to use it in the design and development of large-scale systems. This "fifth discipline" [3] has helped us to puzzle our way through complex problems and improve our ways of thinking, in general (such as "thinking outside the box").

Systems Thinking

At this point, we look more deeply at this important notion (the last point on our list of items on the "systems approach"), and how it has played a role in developing this system architecting method.

A review of several sources dealing with "systems thinking" has led to the following list of features of such an approach [2, p. 32]:

- Holistic
- Integrated
- System wide
- Inclusive
- Expansive
- Fusion
- Top level
- Broader
- Lateral

The last item on this list, namely "lateral," has been defined and set forth in considerable detail by Edward deBono [4], quite a few years ago. Briefly, this means that instead of digging more deeply in one area of investigation it might be better to move over (laterally) and establish one or more new areas of investigation. Also, it might well mean that your current area of investigation might benefit from becoming "broader" rather than "deeper." Two ways in which these notions have been used in formulating the notion for system architecting in this treatise have to do with functional decomposition and the consideration, from the beginning, of alternative architectures. Perhaps this helps to define a new heuristic, something like "when at an impasse, look sideways instead of down."

Complexity

We acknowledge that systems appear to be getting more and more complex. Here are some system features that support these increases in complexity [5]:

- Size
- Modes of Operation

- Nonlinear Behavior
- Human/Machine Interaction
- Functionality
- Duty Cycle
- Degree of Integrationce
- Number/Type Interfaces
- Parallel vs. Serial Operation
- Real Time Operations
- Very High Performance

In the light of this fact, our architecting method needs to be especially suitable to increases in size and functionality. It can become more expansive, as long as it does so linearly rather than exponentially. It also needs to be responsive to the Rechtin suggestion (heuristic) that, where possible, "simplify, simplify, simplify" as well as "KISS" (keep it simple, stupid) [6]. And while we're thinking along these lines, it makes sense to pay attention to Occam's Razor. Rechtin has explained this perspective [6] as: "The simplest solution is usually the correct one."

Human Judgments

As we consider how to architect a system, we spend some time here looking at the human element, in particular within the context of the architecting and the fact that architecting teams are the rule rather than the exception. The context includes proposal evaluations, often by government evaluation teams. These teams need to follow government rules and regulations known as the acquisition system.

The architecting teams need to be highly productive groups, with leaders who know how to get the most out of their teams. They also need to know how to account for the possible biases that various team members might have. During this author's work with the Aviation Advisory Commission (AAC) some years back (see note in Chapter 1), these biases were actually "measured" and used as part of the architecting process [7]. This was done by constructing a matrix of weights for the individual Commissioners. These were specific estimates of how the Commissioners would weigh the various evaluation criteria. These estimates reflect human inputs in the evaluation process. We must not forget that whatever the procedure, it is people that determine how that procedure is applied and provide critical inputs. To be more specific about the AAC experience, the criteria used by the commissioners were:

- Social
- Environmental

- Service quality
- System capacity
- Human factors
- International economic
- Investment costs
- Operating costs.

The weights ranged from 5 percent to 40 percent. This kind of example illustrates how various people can look at the world and come to quite different answers and conclusions.

References

1. NASA Systems Engineering Handbook, NASA/SP-2007-6105, Rev. 1, Washington, DC (December 2007), p. 276.
2. Eisner, H., *Topics in Systems*, Mercury Learning and Information, 2013.
3. Senge, P., *The Fifth Discipline*, Doubleday/Currency, 1990.
4. deBono, E., *The Use of Lateral Thinking*, Pelican Books, 1971.
5. Eisner, H., *Managing Complex Systems – Thinking Outside the Box*, John Wiley, 2005.
6. Rechtin, E., *Systems Architecting*, Prentice-Hall, 1991.
7. Eisner, H., *Computer-Aided Systems Engineering*, Prentice-Hall, 1988, p. 352.

chapter three

What is an Architecture?

Introduction

Chapter 1 closes with the top-level steps of the process of developing an architecture, namely:

1. Functional Decomposition
2. Design Approaches to Instantiate All Functions and Subfunctions (Synthesis)
3. Evaluation of Alternatives (Analysis)
4. Selection of Preferred Alternative (Cost-Effectiveness Assessment).

It is the second step that defines alternative architectures, and at the end of the fourth step, one has selected a preferred architecture from among the alternatives. In all cases, architectures are definitively built upon the functional decomposition of the system.

A Top-Level View

We can get a better idea as to the matter of what defines an architecture by looking at the functional decomposition of a communications system, as illustrated in Table 3.1 [1].

For each of the system functions, one considers alternative design selections, DIJ (design J for function I). All the various combinations represent alternative architectures, by definition. Therefore, the total number of possible architectures, in principle, are (2)(2)(3)(3)(2)(2)(3)(2) or 864 alternatives. This is the concept for developing and considering a set of alternative architectures. As a practical matter, of course, this number is too large. We soon wish to look at a limited number of architectures, of the order of three or four. This notion is explored in more detail below along with a formal definition of the essence of an architecture.

Definition – In Words

Based upon the above concept for developing an architecture, we now formally define it as [1]: "An *Architecture* is an organized top-down selection and description of design choices for all the important system functions

Table 3.1 Functions and Design Choices for a Communications System

Functions	Alternative Design Choices
1. Multiplexing/Demux	D11, D12
2. Modulation/Demod	D21, D22
3. Switching and Routing	D31, D32, D33
4. Encryption/Decryption	D41, D42, D43
5. Formatting/Signal Conversion	D51, D52
6. Control and Monitoring	D61, D62
7. Recording and Playback	D71, D72, D73
8. Satellite/Terrestrial Communications	D81, D82

and subfunctions, placed in a context to ensure interoperability and satisfaction of final system requirements."

Cost-Effectiveness of Systems

When we look at a large number of systems, and in particular their costs and effectiveness, we find that they tend to follow a curve such as that shown in Figure 3.1.

For low values of cost and effectiveness, there is a mostly linear region where both effectiveness and cost both increase together. As both increase, there is a "knee-of-the-curve" behavior as the curve bends to the right. From there, costs increase dramatically in order to achieve increasing levels of effectiveness. These three regions suggest that it is useful to define a set of alternative architectures that attempt to represent the regions. Thus, the choice is made that the architect, or the architectural team, will try to examine three potentially competing alternatives.

System Architecture – Example

The following example of alternative architectures was produced by Richard C. Anderson in one of Professor Eisner's classes on Systems Engineering at the George Washington University. It is a favorite in that it is simple and easy to understand, and illustrates the principles of developing a system architecture. The functional decomposition for the system, known as SCAS (*Severe Climate Anemometry System*), is shown below [1].

Function 1 – Atmospheric Sensing
Subfunction 1.1 – Wind Speed Sensing
Subfunction 1.2 – Wind Direction Sensing
Subfunction 1.3 – Pressure Sensing

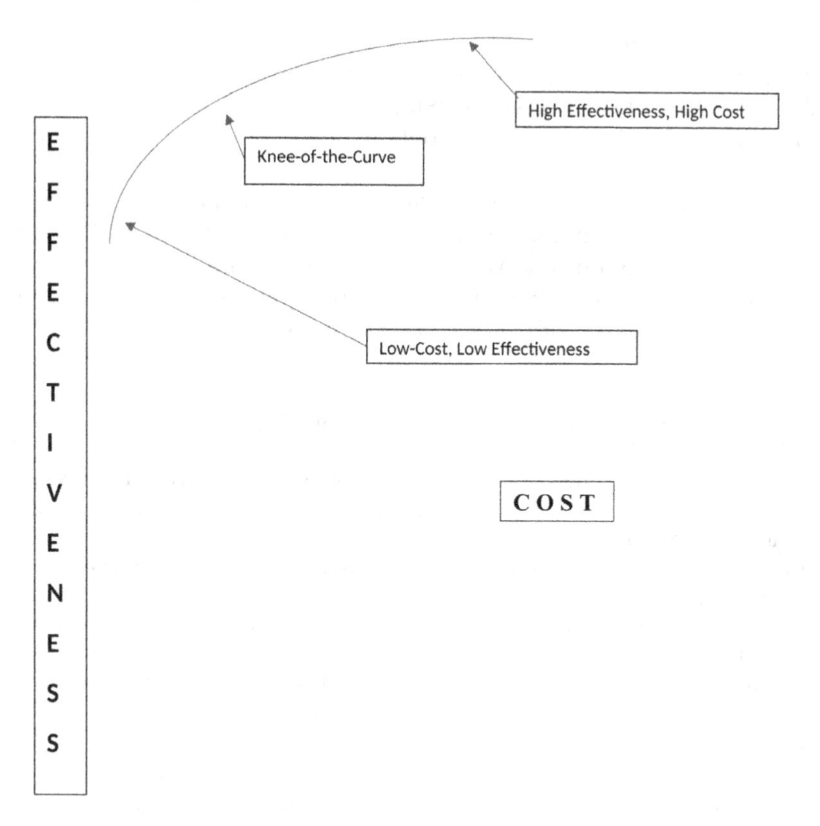

Figure 3.1 Cost-Effectiveness Patterns Over Many Systems.

Function 2 – Mechanical Service
Subfunction 2.1 – Instrument Housing
Subfunction 2.2 – Orientation/Position

Function 3 – Environmental Service
Subfunction 3.1 – Ice Control

Function 4 – Power Service
Subfunction 4.1 – Main Power Supply
Subfunction 4.2 – Power Regulation
Subfunction 4.3 – Backup Power

Function 5 – Indoor/Outdoor Transmission
Subfunction 5.1 – Power Transmission
Subfunction 5.2 – Signal Transmission
Subfunction 5.3 – Physical Linkages

Function 6 – Data Handling
Subfunction 6.1 – Data Collection
Subfunction 6.2 – Data Processing/Storage
Subfunction 6.3 – Reporting/Distribution

Based upon this functional decomposition, three architectures are developed, all three satisfying the basic requirements for the system. They are based upon the notion of looking for a low-cost, knee-of-the-curve, and high-effectiveness set of alternatives. They are formulated with increasing levels of effectiveness, with whatever the resultant costs may be.

Table 3.2 Design Choices for Various Subfunctions

Functions/ Subfunctions	Low Cost	"Knee-of-Curve"	High Effectiveness
1. Atmospheric Sensing			
1.1 Wind Speed Sensing	COTS Pitot Tube	Pitot Tube With Transducer	Add Radio Transducer
1.2 Wind Direction Sensing	Simple Shaft Drive	Simple Shaft Drive	Simple Shaft Drive
1.3 Pressure Sensing	COTS Pitot Tube	Pitot Tube w/ Xducer	Add Radio Xducer
2. Mechanical Service			
2.1 Instrument Housing	Machined Aluminum	Add Molded Composites	Less Weight/ Compact
2.2 Orientation/ Position	Wind-vaned Bearing	Less Tail Boom Length	High Precision Bearing
3. Environmental Service			
3.1 Ice Control	Analog Feedback Temperature Control	Add Digitized Control	Add Process and Heat Pipes
4. Power Service			
4.1 Main Power Supply	Commercial 220/110	Commercial 220/110	Commercial 220/110
4.2 Power Regulation	Conditioners/Rods	Add Ground Fault Interruptor	Add Lightning Arrestor
4.3 Backup Power	Battery Instruments	Gas Generator With Sensor	Hi-Rel Diesel with Switch
5. Indoor/Outdoor Transmission			
5.1 Power Transmission	Stranded Wire Harness	Stranded Wire Harness	Custom Slip Rings
5.2 Signal Transmission	Foil-Shielded Wire	Coax w/Slip Rings	2-Way Radio, no wiring
5.3 Physical Linkages	Shaft/Conduit	Add Shielded Xducer	Minimum Shaft

Continued

Functions/ Subfunctions	Low Cost	"Knee-of-Curve"	High Effectiveness
6. Data Handling			
6.1 Data Collection	Potential and Indoor Pneumatic Cell	Magnetic Position Sensor	Optical Position Sensor
6.2 Data Processing/ Storage	Manual Database Entry	Automatic Computer Control	Automatic Computer Control
6.3 Reporting/ Distribution	Physical Manual	GUI + Modem Access	DBMS + Packet Network

With respect to these architectures, we make the following overall observations:

1. Looking at each architecture, we check to verify that the design choices are interoperable by examining each item in each column, against all other items in that column.
2. We note that the "Low Cost" alternative is designed to satisfy all of the firm requirements for the system, and that the other alternatives represent increases in system effectiveness (higher-performing, generally higher cost).
3. We note also that in some cases, the design choices may be the same (e.g., the "stranded wire harness" for the "Low Cost" and the "Knee-of-Curve," which changes to the "Custom Slip Rings" for the "High Effectiveness" architecture; the "wind direction" is measured, in all three architectures, by a simple shaft drive).
4. The given architectures are developed using the "KISS" principle (keep it simple, stupid) [2], with considerable hidden (but implicit) information about each alternative.
5. The architectures are assumed to be formulated by a team of "architects" such that the information is produced by a group rather than a single individual.
6. The team of architects reserves the right to modify any and all entries if the interactions between team members so indicate.
7. The synthesis contains all the functions that the system is to contain, and that no functions are missing. Otherwise, we have an incomplete architecture which will likely lead to poor architectural constructs.
8. At this point in the architecting process, we have three competing alternatives, and have not yet selected a preferred alternative.
9. The general name for developing the given data is to engage in a "synthesis" procedure.

10. The three architectures are formulated by using the collective experience, judgments, and intuition of the architecting team.
11. The overall architecting process is compatible with a four-week completion time period, with increases in detail and fidelity over longer time frames (to include many months).

In the next chapter, we move on to present a more complete procedure that articulates the "short-form" method of analysis of the three alternatives so as to proceed in the direction of a preferred alternative.

References

1. Eisner, H., *Essentials of Project and Systems Engineering Management*, 3rd Edition, John Wiley, 2008; the architecting of the SCAS system was originally carried out as a class assignment by Richard C. Anderson (now Dr. Anderson).
2. Rechtin, E., *System Architecting*, Prentice-Hall, 1991.

chapter four

Evaluation of Alternatives

Introduction

The next step in the process of architecting is to formally evaluate the alternatives that have been defined in the "synthesis" step. The main basis for such an evaluation is to compute the cost and the effectiveness of these alternatives and then compare them. The critical aspect of this calculation, in terms of basic concept, is to confirm how effectiveness is to be measured. Once this is done, we are in a position to say that cost-effectiveness will be the key component of the evaluation.

However, even after a careful cost-effectiveness evaluation, there are still some factors that need to be considered before a preferred architecture is selected. These factors have to do with the ground-rules under which the system in question is being acquired.

System Effectiveness

In the previous chapter, three alternative architectures for an anemometry system were defined (synthesized). For this particular system, we will now proceed to establish five criteria that will form the basis for measuring the effectiveness of this system. These criteria are:

1. Performance
2. Human factors
3. Reliability
4. Maintainability
5. Risk.

We acknowledge that these criteria may have different levels of importance and so we assign a set of weights (w_i) to them. Then, we will use a classical rating scheme to perform the effectiveness calculations. Each alternative is rated against each criterion, the ratings are weighted, and then the sums represent the effectiveness measures.

This well-known weighting and rating procedure is particularly well-suited to the demands of architecting within severe time constraints (less than one month). We do not have time to do an in-depth analysis but do have enough time to carry out this process in a group setting. Such a setting allows for intensive group discussions and interchange of ideas and viewpoints. The bottom line is that we are able to calculate meaningful measures of effectiveness, in a team context, and move on to the next steps that ultimately will lead to a preferred architecture.

Using the example from the previous chapter, we calculate the numerics below to illustrate the procedure. This represents an "evaluation framework" and leads to measures of effectiveness for each architecture. The simple calculation is below, along with the framework

$$\text{Effectiveness (architecture j)} = \sum wt_{ij}\, r_{ij}$$

where wt_{ij} is the weight of the *i*th criterion and r_{ij} is the rating against that criterion.

Cost estimates are added, based upon standard cost estimating procedures, as described in a later chapter (see Table 4.1).

The next step involves placing these numbers in a graphic format, as shown in Figure 4.1.

Commentary on Graph of Costs and Effectiveness

First, we note that the plot of Figure 4.1 has the same general shape of Figure 3.1 in Chapter 3. That is, each of the three architectures has increasing effectiveness as well as costs in relation to the others. The shape has the familiar "bending over" after the knee-of-the-curve solution.

Second, let us focus on the low-cost alternative. It has the lowest effectiveness score and also the lowest cost, as expected. We note that if we have only $1.2 million to spend, then we are driven to this system, even though we may wish to go for another solution. This is known as a cost-constrained solution, and is quite common in today's world. Many

Table 4.1 Effectiveness and Cost Measures

Evaluation Criteria	Weights	System Under Consideration					
		Low Cost		Knee-of-Curve		High Effectiveness	
		Score	Wtx Rating	Score	Wtx Rating	Score	Wtx Rating
Performance	30%	6	1.8	7	2.1	9	2.7
Human Factors	20%	8	1.6	8	1.6	8	1.6
Reliability	15%	7	1.05	8	1.6	9	1.35
Maintainability	20%	8	1.6	8	1.6	9	1.8
Risk	15%	7	1.05	9	1.35	9	1.35
SUMS (MOEs)	100%	–	7.1	–	8.25	–	8.8
COSTS	–	–	$1.2M	–	$1.4M	–	$1.8M

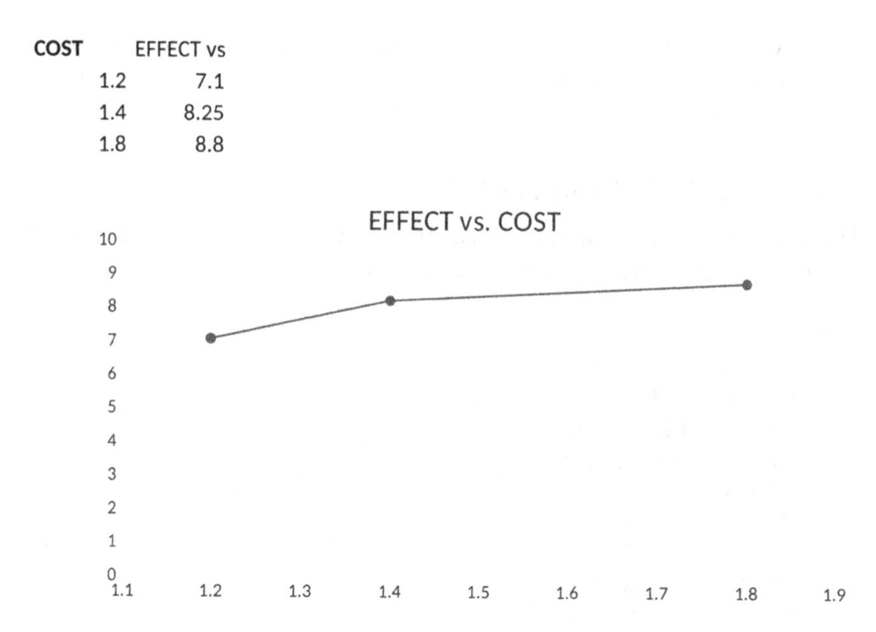

COST	EFFECT vs
1.2	7.1
1.4	8.25
1.8	8.8

Figure 4.1 Effectiveness and Costs for Three Anemometry Architectures (Effectiveness on scale of 1–10, Costs in Millions of Dollars).

purchasers will go for the low-cost approach as long as it satisfies the requirements. In that sense, the alternatives are not really competitors with respect to one another. But we need to perform the calculations in order to make this discovery.

Next, we move on to the "knee-of-the-curve" architecture. We are not really sure that this is the true knee-of-the-curve, but we do observe a significant increase in effectiveness above the low-cost system. This alternative is generalized to represent the "best value" system in that we gain increasing effectiveness for modest increases in cost. In today's world of acquisition and procurement, this can be a quite good solution, if the "rules" permit it as the preferred answer. "Best Value" has a definite attraction, and for good reason. Of course, we need to be able to afford the increased cost, which has moved from the low cost of $1.2 million to $1.4 million.

We then look at the third architecture, the high-effectiveness system. We see the "bending" represented by a modest increase in effectiveness (to 8.8) with a relatively large increase in cost (to $1.8 million, from $1.4 million). This is the best performing system, at least in terms of "paper" design and intention. If we have the funds, we may wish to go for such a system. However, acquisition rules tend not to support such a choice, except under special circumstances, like certain military systems.

So we see that the three alternatives are not "pure" competitors; there are several other considerations when it comes to picking a preferred system. More will be said about this notion in Chapter 8.

Additional Fidelity In Rating

The various architects might benefit from additional fidelity in the rating scheme. For this purpose, we define further detail as illustrated below.

1. Performance
   1.1 Vaning Stability
   1.2 Average Power Consumption
   1.3 Impact Resistance
   1.4 Data Availability
   1.5 Useful Life
2. Human Factors
   2.1 Ease of Use
   2.2 Operator Safety
   2.3 Bystander Safety
3. Reliability
   3.1 Generic Failure Rate
   3.2 Level of Redundancy
   3.3 Basic Design Structure
4. Maintainability
   4.1 Frequency of Scheduled Maintenance
   4.2 Ease of Maintenance
   4.3 Complexity of Assembly
5. Risk
   5.1 Schedule Risk
   5.2 Performance Risk
   5.3 Cost Risk.

This additional detail can be used formally or otherwise as an aid in providing the "rating" estimates. A formal approach would be to assign weights and develop ratings for each and every sub-criterion.

Other Factors

The ratings for the various criteria are shown here on a scale of 1–10. The meaning of each numeric needs to be explained, in words, so that the evaluations become more uniform and consistent among the evaluators. This can be done at a meeting of the team of architects or by an individual with expertise in this area. An alternative is to move to a college scoring system from A to F, with obvious and well-known numeric equivalence. In any case, we are interested in obtaining the best results among evaluators.

We noted as well, in Chapter 1, the presence of bias among the evaluators. This was demonstrated by a matrix of weights for each criterion, for the evaluators. This approach might well be used in order to obtain the overall weights (averages) for these criteria. This will tend to improve the overall process when there are several evaluators, as with a team of architects.

The evaluation process is also improved by several changes in procedure, otherwise known as a set of sensitivity analyses. Here we are looking for variations in the results when changes are made in the weighting factors, the rating schema, adding new evaluation criteria, and others as suggested by members of the architecting team. Sensitivity analysis is a well-known process used to explore variations in important variables, factors, and results.

The evaluation process of weighting and rating has been examined in quite a lot of detail under the category of Multi-Attribute Decision Making (MADM). Examples are provided in the last three references [1, 2, 3] listed below.

Closing Thought

We also take note here of the context for developing architectures for systems. There are two over-riding considerations. The first is that we are a doing preliminary design for a system, and a key part of that design is the formulation of a system architecture. The second is that we are on a short time-line, of the order of a month or two. Therefore, we must rely in good measure on subjectivity and the collective experience of a team of system architects. We presume that we will have time, later in the system development cycle, to examine more detailed issues to include system behavior (performance) and tradeoffs.

References

1. Hwang, C. and K. Yoon, *Multi-Attribute Decision Making*, Springer, 1981.
2. Greco, S., M. Ehrgott, and J. Figuiera (Eds.), *Multi-Criteria Decision Analysis*, Springer, 2005.
3. Gibson, J., W. Scherer, and W. Gibson, *How To Do Systems Analysis*, John Wiley, 2007.

chapter five

Architecting a House

Introduction

In this chapter we develop another illustrative example of the architecting process, in this case a standard residence. We see that the overall architecting process is applicable to the domain of a house, and explore what that might entail.

We start out with the same first step, namely, a functional decomposition of the residence. Then we move on to the other well-defined steps in the overall procedure.

Functional Decomposition

We formulate the functional decomposition, as shown below.

1. Environment (Heat, AC)
2. Sleeping Quarters
3. Food Preparation
4. Bath Areas
5. Style/Design
6. Recreation (Deck, Pool)
7. Space/Size
8. Living/Dining Space
9. Security
10. Lawn/Garden
11. Car Facility
12. Plumbing
13. Electrical
14. Special Amenities.

Synthesis of Alternative Architectures

As with the previous example of an anemometry system, we next develop three alternative architectures, as provided in Table 5.1 (including cost estimates for each of the architectures).

We note that in one of the entries (Item 7), we cite numbers that are representative of the function. In effect, we are specifying parameter "targets" for space/size.

Table 5.1 Functions and Costs for Home Systems

Functions	Low Cost	Knee-of-Curve	High Effectiveness
1. Environment	Single Furnace Low Capacity A/C	Two Zone Furnace Medium Capacity A/C	Three Zone Furnace High Capacity A/C
2. Sleeping Quarters	Three Bedrooms	Four Bedrooms	Five Bedrooms + Den
3. Food Preparation	Standard Kitchen	Standard + Hi Xtras	Islandt + Xtra Counters
4. Bath Areas	Standard Toilet and Bath	Add Sinks and Bidet	Add Closets and Jet Bathtub
5. Style/Design	Ranch Style	Faux Farm	Contemporary
6. Recreation	Small Deck	Front and Back Decks/Porches	Wraparound Decks + Pool
7. Space/Size	3,500 ft^2	4,500 ft^2	6,500 ft^2
8. Living/Dining Space	Standard Spaces	Large Spaces	Very Large Spaces and High Ceilings
9. Security	No Extra Security	Camera/Tape	Add Alarms and Internet
10. Lawn/Garden	Small Lawn	Buried Water Lines	Large Lawn/ Gazebo
11. Car/Facility	One Car Garage	Two Car Garage	Three Car Garage
12. Plumbing	Standard Copper	Multiple Flareouts	Add Sprinkler System
13. Electrical	Standard Amps	Add 50% More	Add 100% for Growth
14. Amenities	Standard Closets	Add Built-in Book Cases	Add Library/ Video Room, and Elevator
COST	$700,000	$1,200,000	$2,500,000

Evaluation Framework

We next set forth an evaluation framework, to include evaluation criteria and a weighting and rating procedure. Perhaps the most difficult part of this step is to decide upon the set of evaluation criteria which have been selected as:

1. Performance
2. Maintainability
3. Holding Value

Table 5.2 Effectiveness and Costs for Home Systems

Evaluation Criteria	Weights	Low Cost		Knee-of-Curve		High Effectiveness	
		Score	Wtx Rating	Score	Wtx Rating	Score	Wtx Rating
Performance	30%	6	1.8	8	2.4	10	3
Maintainability	25%	5	1.25	7	1.75	9	2.25
Holds Value	10%	6	0.6	8	0.8	9	0.9
Risk	15%	6	0.9	7	1.05	8	1.25
Feels Like Home	20%	6	1.2	8	1.6	9	1.8
SUMS (MOEs)	100	–	5.75	–	7.6	–	9.2
COSTS	–	–	$700K	–	$1,200K	–	$2,500K

(System Under Consideration: Low Cost, Knee-of-Curve, High Effectiveness)

4. Risk
5. Feels Like Home (see Table 5.2).

The overall effectiveness numbers as well as the costs are shown in the cost-effectiveness graphic Figure 5.1.

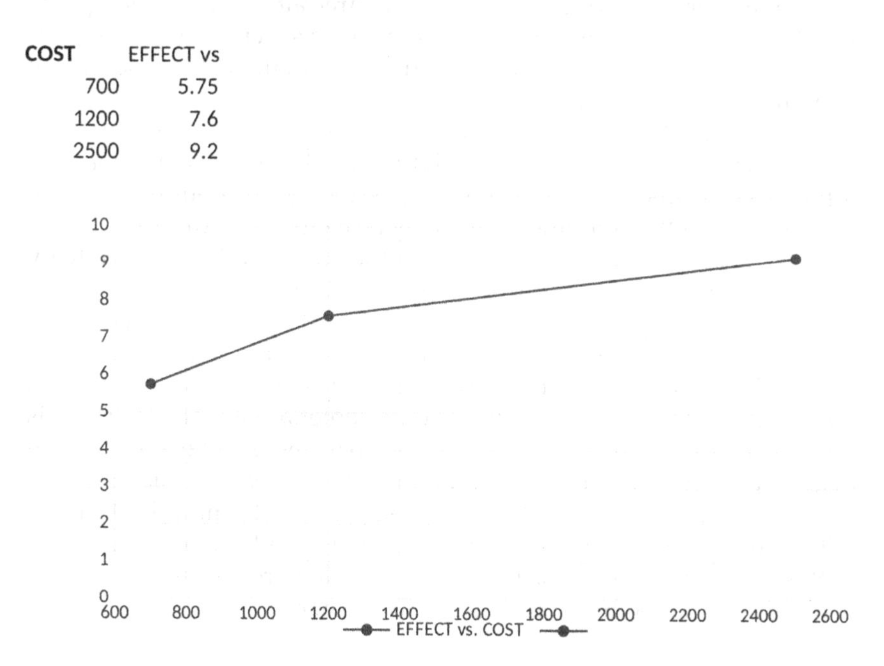

Figure 5.1 Effectiveness and Costs for Three Home Architectures.

Commentary – Functional Decomposition

We take a hard look at the functional decomposition and notice that there is a strong attempt at citing functions rather than more familiar parts of a house. For example, instead of talking about a kitchen, we use the phrase "food preparation." In some cases, there is some ambiguity that will be resolved when the function is instantiated. Such is the case with "style/design" and also "special amenities." The first listed function of environment is possibly not clear and it also might be broken down into its subordinate functions of heating and air conditioning. Further, it may be that there are other functions not addressed, but that is for the architect to discover. Typically, for real large-scale systems, there is a requirements document that defines all the functions to be part of the system.

Commentary – Synthesis

Here we note that each of the 14 functions has been instantiated by a design choice, at some appropriate level of detail, and for three alternatives. There is considerable freedom to specify these design choices. For example, the choices for the "environment" function stop short of specifying the air conditioning capacity but make it clear that we wish to have a multiple zone system for two of the three alternatives. For "space/size," we note that the number of square feet is specified. Perhaps there is a better way to deal with this "function," as might be the case as well with the entries for "living/dining space."

We see similar choices, for example, with the "electrical" function (specific amperage not specified) and the special amenities function. The differences between the three alternatives can be seen but the numerical values are left unspecified. This may be changed by the architecting team if it wishes to be more specific. On the other hand, such specificity may be considered to be a future step in the design process.

We note also that the three alternatives are based upon three styles, namely, ranch, faux farm, and contemporary. This is one way of making choices. However, it may be that the architecting team does not wish to do this. Possibly it wants to consider only contemporary styles. If that is the case, there is no need to consider the ranch and faux farm styles – all three alternatives would be based upon a contemporary design style.

To complete the top-level architecting process of synthesis, the costs of the three alternatives are either calculated or indeed are "specified." In this example, they are part of the process and are necessary in terms of deciding what architecture ultimately to choose.

Commentary – Evaluation Framework (Analysis)

This third item in the process has four steps:

1. Defining the evaluation criteria
2. Developing the weights for the criteria
3. Formulating the ratings for each criterion
4. Carrying out the effectiveness calculations (weighting and rating).

As shown earlier in this chapter, five overall criteria were established, i.e., (1) performance, (2) maintainability, (3) holding value, (4) risk, (5) feels like home. After applying the weights, the overall effectiveness numbers were:

Low-Cost architecture – 5.75
Knee-of-the-Curve – 7.6
High Effectiveness – 9.2

If these values are placed on a graph, along with the costs ($700K, $1,200K, and $2,500K) as the x-axis, we see that these "points" are consistent with representations described earlier.

It would seem that cost is a major driver in thinking about a preferred architecture, from a consumer's point of view. Folks who have lots of money might well choose the high-effectiveness architecture just because they can afford it. Likewise, if cost is a problem, the low-cost alternative is likely to be chosen. And the knee-of-the curve architecture is attractive for those with more available assets. In all cases, individual choices depend strongly upon what the consumer wants, including tradeoffs that are preferred by the consumer.

So we see here a second example of the overall process, developed to be as simple as possible (remember KISS – keep it simple, stupid), and designed to show comparisons between alternative architectures. The format of the "synthesis" step is itself indicative of how an architecture is to be represented. Indeed, this step is simple, but very specific in what it is and what it depicts. It also shows the key features and the key tradeoffs that need to be considered.

chapter six

Architecting an Automobile

Introduction

In this chapter we proceed with another illustration – that of architecting an automobile. We will follow the essential steps of the previous two chapters. This example will further demonstrate the method of architecting set forth in this treatise.

Functional Decomposition

We begin by articulating a functional decomposition, as shown below.

1. Speed Control
2. Braking
3. Carrying Capacity (Pax, Cargo)
4. Styling/Appointments
5. Automation
6. Visibility/Human Factors
7. Safety/Security
8. Power Source/Supply
9. Audio/Video
10. Emission Control
11. Suspension
12. Maintenance
13. Fuel Economy

Synthesis of Three Alternative Architectures

Based upon this decomposition, we next define three alternative architectures for the automobile. These are shown in Table 6.1. In this case there are some 13 functions to be considered.

Evaluation Framework

The next step is to define a set of evaluation criteria, which will be taken to be:

1. Performance
2. Maintainability

Table 6.1 Functions for Alternative Automobile Systems

Functions	Low Cost	Knee-of-Curve	High Effectiveness
1. Speed Control (0–60)	6 Second Standard	5 Second Additional Beeps	3 Second Beeps + Visual
2. Braking (from 60 mph)	200 ft	160 ft	120 ft
3. Carrying Capacity	5 pax 30 cubft	5 pax 34 cubft	5 pax 37 cubft
4. Styling	Vinyl	Vinyl + Leather	Leather
5. Automation	None	Driver Assist	No Hands Driving
6. Visibility/Human	Standard	Add Special Mirrors	Add Video Assist
7. Safety/Security	Standard	Standard + Special Mirrors	Add Hi-Performance Airbags
8. Power Source	4 Cylinder/Gas	4 Cylinder/Gas	6 Cylinder/Gas
9. Audio/Video	Beeps	Sound + Video	Sound + Video
10. Emission Control	Standard	Standard	Hi-Performance Control
11. Suspension	Standard	Standard	Hi-Performance
12. Maintenance	Dash Display Special Codes	Augmented Dash Special Codes	Augmented Dash AI Special Codes
13. Fuel Economy	20 mpg	22 mpg	26 mpg

3. Holds Value
4. Reliability
5. Human Factors
6. Overall Aesthetic.

We note the persistence of "performance" and "maintainability" as criteria for several of the systems we are considering. This is not unusual for many types of systems. Sub-criteria help to establish the factors that go into the meaning of these notions. Such sub-criteria were given as examples in the architecting of the anemometry system in Chapter 4.

We next proceed with evaluating the three alternatives, with the weighting and rating procedure that we have been using. The framework is provided in Table 6.2.

Table 6.2 Effectiveness and Cost Measures for Automobile Systems

Evaluation Criteria	Weights	System Under Consideration					
		Low Cost		Knee-of-Curve		High Effectiveness	
		Score	Wtx Rating	Score	Wtx Rating	Score	Wtx Rating
Performance	30%	6	1.8	8	2.4	9	2.7
Maintainability	20%	8	1.6	9	1.8	9	1.8
Holds Value	15%	6	0.9	8	1.2	9	1.35
Reliability	15%	8	1.2	9	1.35	10	1.5
Human Factors	10%	7	0.7	9	0.9	10	1.0
Overall Aesthetic	10%	7	0.7	8	0.8	9	0.9
SUMS (MOEs)	100%	–	6.9	–	8.45	–	9.25
COSTS	–	–	$35K	–	$50K	–	$85K

We proceed by looking at these numbers on a graph, as shown in Figure 6.1

COST	EFFECT vs
35	6.9
50	8.45
85	9.25

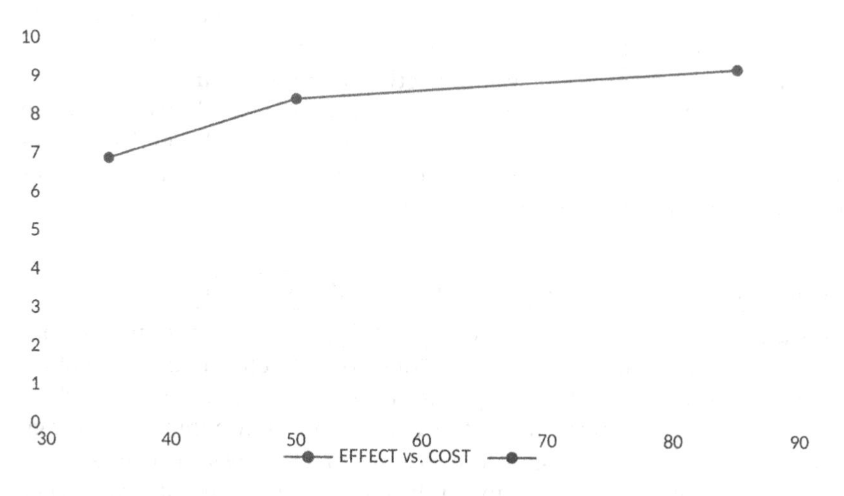

Figure 6.1 Effectiveness and Costs for Three Anemometry Architectures (Effectiveness on scale of 1–10, Costs in Millions of Dollars).

Commentary – Functional Decomposition

As with the architecting of a house in the previous chapter, some of the functions are represented by a set of numbers, a set of specifications. Such is the case with speed control (time to go from zero to 60 mph), braking distance (feet), and fuel economy (miles per gallon). There are a total of some 13 functions, and the reader may challenge this selection by adding functions or being more precise about what the function is, as for example with "automation" and possibly "safety/security." If the architecting is carried out in a group or team setting, it is expected that a set of functions will ultimately be agreed upon. One can imagine that a team of architects might be assembled in an automobile company to consider a "line" of vehicles within an overall category of vehicles.

Commentary – Synthesis

The synthesis step is, of course, the critical part of the architecting process. We see the typical breakdown into three architectures – low cost, knee-of-the curve, and high effectiveness. As indicated in previous chapters, this is the "heart" of the process and Figure 6.1 represents and defines what an architecture is, in very concrete terms. As suggested above, some of the functions are specified by numbers, and do not suggest a specific way in which to achieve those numbers. Where possible, the architect might well consider how this challenge could be met. For example, with the braking function, alternative design approaches might be considered which will meet the specifications shown for each alternative. In some cases, the approach to the low-cost and the knee-of-the curve architectures is the same. This is true for "carrying capacity, safety/security, emission control and suspension." This does not mean that such an approach cannot be challenged and changed. The architecting team, of course, can modify this approach for good and sufficient reasons. In general, however, we see improvements in performance in moving from left to right on the synthesis chart. This is why we expect the evaluations to reflect this improvement in the next step (analysis) of the overall process.

Commentary – Evaluation Framework (Analysis)

As is the case for architecting a house in the previous chapter, the evaluation framework involves four important steps: (1) defining the evaluation criteria, (2) establishing the weights for each of the criteria, (3) providing the ratings, and (4) performing the calculation of weights and ratings (products and sums) to obtain the overall effectiveness measures. Here again, we maintain the separation of the alternatives as an integral part of the evaluation process.

The bottom line numbers reveal the pattern that we thought we might see – a significant increase in effectiveness at little extra cost in going from the low cost to the knee-of-the curve, and a significant increase in cost, with not much increase in effectiveness in moving from the knee-of-the curve to the high-effectiveness architecture. There is no guarantee that such would be the case, but we tend to appreciate that we are likely to have a knee-of-the curve alternative when we see this type of pattern.

If the architect or the architecting team is not pleased with the results of this step in the process, it should be repeated and specific sensitivities should be explored. For example, what changes occur in response to changes in the weighting factors? What changes occur in response to a shift from a zero to ten rating system to a change to a college scoring system (A, B, C,...)? What changes occur in response to a change in the evaluation criteria? What changes occur when there is a procedural change in the personnel that are carrying out the evaluation step? Do we get a different "answer" when the team composition is modified, or is there quite a lot of agreement even as this team is changed?

If we place this architecting process into a setting inside an automobile company, or a "consumer reports" type of company, we might well obtain some additional insight into what the additional features cost, and what a consumer (customer) might be willing to pay for such features. In other words, what might the prices be as we move from the low-cost to the high-effectiveness alternatives?

Above all, as expressed previously, the suggested architecting process is simple and specific. It relies upon these features in order to develop specific results in a relatively short period of time. It depends upon the explicit consideration of alternatives in distinction to early definition of a preferred architecture which might be wrong. Rather than emphasizing the process of creating more and more data (drilling down) regarding a single alternative, it relies on a process of "lateral thinking" as we bring alternatives into consideration. It relies on a procedure that has gained some traction in the Department of Defense known as "analysis of alternatives (AoA)" [1]. But entering that world involves at least one other treatise.

Reference

1."Analysis of Alternatives (AoA) Handbook," see www.prim.osd.mil.

chapter seven

Commentary

A Preferred Architecture

Introduction

The last several chapters have dealt with the matter of architecting sets of alternatives. These have been low-cost, knee-of-the-curve, and high-effectiveness architectures. That is, we have attempted to construct alternatives in these regions or domains. Readers may look at these and react by selecting the one architecture they appear to prefer. Or they may wish to explore the matter in greater detail to find a preferred alternative. In general, additional data as well as thought will be part of selecting a preferred architecture. Some of those considerations are examined in this chapter.

Top Level Considerations

We start with the notion that we have representations of a low-cost, knee-of-the-curve, and high-effectiveness architectures.

The low-cost alternative will become the preferred selection if we do not have the funding that will allow us to move beyond this choice. A low-cost system is highly attractive for many procurements, for obvious reasons. Indeed, given that the system meets all the requirements, some acquisition agents will insist that this is the proper choice. Therefore, there is great pressure to elevate the low-cost system to "preferred" status.

The knee-of-the-curve alternative is also very attractive and is often referred to as the "best value" selection. We are achieving large increases in effectiveness for modest increases in cost. This is happening just as the curve bends and requires quite substantial increases in cost. System acquisition agents will tend to support the notion that the "best value" approach is preferred. Of course, one needs to be able to support that choice by putting up the appropriate funding.

The high-effectiveness alternative tends to be best for those who have the funds to support such a choice. They wish to have the benefit of very high effectiveness and performance. They generally are willing to pay for "extras." They wish to have special amenities and find that cost is not an obstacle to getting what they want.

Another context for having the high effectiveness alternative become the preferred architecture is that of military procurements. The notion here is that those ordering military systems do not wish to have our country go into battle with less than the best performing systems. We will not have an F-35 joint strike fighter that is other than the absolute best we are able to conceive of and build. This, of course, makes sense and accounts for some of the high costs that have been part of such programs.

Finally, if we are to select a preferred architecture, we need to understand all of the procurement and acquisition "rules" that will be used to award contracts. Industry responds to these rules, and makes decisions based upon strong motivations to win. For example, if one knows that there is a strong preference for lowest-cost proposals, such may be the basis for making that the preferred architecture and system.

Requirements

In the previous examples of the architecting process, we see little to no mention of requirements. This does not mean they are absent from this process. The ground rule is that all three alternatives absolutely satisfy all of the agreed upon requirements. The architecting team knows this, and adheres to this practice. It is implicit rather than explicit.

Specifications

If there are any extraordinary specifications for the system, they too have been taken into account in carrying out the architecting procedure.

Minimal Explicit Information

The synthesis tables list the design choices for each of the system functions and subfunctions. This is the absolute essence of the architecting process. Indeed, these are descriptions of what is meant by an architecture. As represented in these tables, they convey a minimum of information. However, each and every entry is supported by back-up data that describe that design choice in detail. The minimal information approach is considered to be in consonance with Rechtin's [1] KISS (keep it simple, stupid) principle. Keeping it simple means that it is better understood by those that are part of and using the architecting procedure. These people have been referred to as the architecting team. More about the operation of a team and the team leader in Appendix A.

Are These Architectures Optimal?

There is no claim that developing architectures with this procedure is optimal. First, we have a short time frame for the overall process (about a month or two). This is not conducive to optimizing anything. Second, the design choices are made by the architecting team so that one might say it is the best that such a team is able to produce within the given time constraint. A senior team that has worked together in the past, several times, is likely to produce a very good set of architectures. In any case, we are saying that the "goodness" of the architectures is clearly related to the team and its degree of experience and competence.

Several (Not One) Architecture

Implicit in the above is the fact that this approach to architecting focuses on alternatives from the beginning. This has the benefit of enlarging the space in which solutions may be found. It increases the likelihood that an appropriate solution will be found.

Assuring Interoperability

The "synthesis" table provides a list of design choices for the various functions and subfunctions. This facilitates a check of interoperability by moving down each vertical column and verifying that these design choices will interoperate with one another. If there is some incompatibility, then one or more design choices need to be changed. Typically, but by no means always, the functions can be instantiated independently of one another.

Views

It is noted that the architecting process is more-or-less independent of the notion of "views." That is to say, architecture views play essentially no part in developing and analyzing the alternative architectures. This is significant since, for example, the Department of Defense Architectural Framework (DoDAF) approach [2] starts out with "views" and somehow that leads to developing a system architecture. Thus we have a fundamental change from the DoDAF approach. To generalize, the notion is that if one starts in the wrong place, it is difficult (if not impossible) to come to the proper solution. The approach here, of course, is clear on the starting point for developing architectures – it is the functional decomposition of the system. Additional discussion on the matter of "views" is provided in Chapter 8.

Architectural Descriptions (ADs)

Another source of "views" is the so-called standard IEEE P1471. We will take some space here (see Chapter 8) to look at what this document says that might be compared with the suggested approach as well as the outputs that this approach produces.

Other Evaluation Criteria

We have set forth a series of evaluation criteria for the three illustrative architectures. These may be expanded in order to have more choices and thus possibly enhance the evaluation process. A more complete list of evaluation criteria [3] that may be used for specific transportation and communications systems is provided below.

Transportation Systems

Availability
Capacity
Capacity-to-Demand Ratio
Comfort and Convenience
Environmental Effects
Frequency of Service
Fuel Consumption
Growth Capability
Maintainability
Quality of Service
Reliability of Service
Resilience
Risk
Safety
Security
Speed
Sustainability
Trip Time

Communications Systems

Availability
Bandwidth
Capacity
Connectivity
Expandability
Grade of Service

Number of Channels (by type of channel)
Quality of Service
Reliability
Response Time
Risk
Security
Speed of Service
Survivability

The shape of the cost-effectiveness curve is, of course, an integral part of selecting a preferred architecture. If, moving from architecture A to architecture B, one is moving from right to left close to vertically, then one has the anomaly that increases in effectiveness are being achieved at little cost. Nonetheless, this may be the case and it points to architecture B being a strong candidate for preferred status. The mature architecting team will go beyond curves and ask many questions about how this anomalous situation came to be. Usually, this also involves a detailed discussion about each architecture in relation to each evaluation criterion. It also likely involves digging more deeply into the question – what is it about alternative B that allows for the strong increase in effectiveness at such a low expenditure?

Other ideas regarding the selection of a preferred architecture can be found in books dealing with decision-making under uncertainty. Special attention should be paid to "Multi-Attribute Decision-Making" (MADM) [4, 5, 6) and similar titles. The reader is advised to stay close to practical real-world considerations and away from purely mathematical constructs.

References

1. Rechtin, E., *System Architecting*, Prentice-Hall, 1991.
2. DoDAF Version 2.02, see dodcio.defense.gov.
3. Eisner, H., *Essentials of Project and Systems Engineering Management*, 3rd Edition, John Wiley, 2008.
4. Hwang, C. and K. Yoon, *Multi-Attribute Decision Making*, Springer, 1981.
5. Greco, S., M. Ehrgott, and J. Figuiera (Eds.), *Multi-Criteria Decision Analysis*, Springer, 2005.
6. Gibson, J., W. Scherer, and W. Gibson, *How To Do Systems Analysis*, John Wiley, 2007.

chapter eight

Descriptions, Views, and Tradeoffs

Introduction

This chapter briefly explores the topics of architecture descriptions, views, and tradeoffs. All are important, with "views" taking a special position due to the initiatives of the Department of Defense in relation to DoD Architectural Framework (DoDAF). This is a surprisingly dense subject that needs to be examined from several "angles." Finally, tradeoffs are an important part of designing any large-scale system and will be addressed mostly by example.

Descriptions

The notion of "architectural descriptions" (ADs) has been part of the scene and has helped in clarifying the issue of "what is an architecture?" Two standards have played an important role, each of which is briefly cited below.

The standard known as P1471 [1] deals with a recommended practice for AD. In that standard, a clear distinction is made between "architectures" and "architectural descriptions." The practice itself deals with the creation, analysis, and sustainment of architectures of software intensive systems. The value and use of ADs are cited explicitly as is the matter of who are the intended users. Here are four interesting quotes from this standard:

- "there has not yet emerged any reliable consensus on a precise definition of a system's 'architecture', how it should be described, what uses such descriptions may serve, or where and when it should be defined"
- "this standard codifies those elements on which there is consensus; specifically the use of multiple views, reusable specifications for models within views, and the relation of architecture to system context"
- "architecture: the fundamental organization of a system embodied in its components, their relationship to each other and to the environment and the principles guiding its design and evolution"

- "the term 'view' is used to refer to the expression of a system's architecture with respect to a particular viewpoint."

This author's bottom line with respect to this useful standard is that it still is not precise in its description of a system's architecture.

Another standard, following the above P1471, is the so-called ISO/IEC 42010 [2]. This standard deals with both systems and software engineering with respect to ADs. It also applies to enterprise architectures. It tends to focus on ADs, architecture frameworks, and description languages. It defines several terms, including:

- Architecting
- Architectures
- Architecture Descriptions
- Architecture Frameworks
- Architecture Views
- Architecture Viewpoints
- Architecture Models
- Architecture Languages
- Roles of Stakeholders.

Here again, despite the specificity of this standard, we do not see a precise definition of an architecture that is commensurate with that set forth in this treatise. Therefore, with respect to this standard, there remains a missing link in what might be the formulation of a method of system architecting. So – in this case – the generality of its approach appears to not work in its favor. That is, the generality does not give us a precise method of architecting a system at the level of detail presented here.

Views

Views of systems and their architectures were cited as important when they were the leading edge of the DoDAF approach to the architecting issue. We recall the three architecture views [3], as cited in Chapter 1:

- The operational architecture view
- The systems architecture view
- The technical architecture view.

The *essential* products as well as the *supporting* products for these views were defined as shown below [4]:

Essential Products
AV-1 – Overview and Summary Information
AV-2 – Integrated Dictionary

OV-1 – High Level Operational Graphic Concept
OV-2 – Operational Node Connectivity Description
OV-3 – Operational Information Exchange Matrix
SV-1 – System Interface Description
TV-1 – Technical Architecture Profile (Standards That Apply)

Supporting Views
OV-4 – Command Relationships Chart
OV-5 – Activity Model
OV6a – Operational Rules Model
OV6b – Operational State Transition Description
OV-7 – Logical Data Model
SV-2 – Systems Communication Description
SV-3 – Systems Matrix
SV-4 – System Functionality Description
SV-5 – Operational Activity to System Function Traceability Matrix
SV-6 – System Information Exchange Matrix
SV-7 – System Performance Parameters Matrix
SV-8 – System Evolution Description
SV-9 – System Technology Forecast
SV-10a – Systems Rules Model
SV-10b – Systems State Transition Description
SV-10c – System Event/Trace Description
SV-11 – Physical Data Model
TV-2 – Standards Technology Forecast

Clearly, a great deal of thought went into the definition of these views. However, it is noteworthy that a specific definition of the system architecture is still missing.

Propositions

There appear to be three propositions with respect to "views," and as defined by this author, that are especially relevant to cite at this time:

1. Accurate and relevant views of an architecture cannot be set forth until the architecture itself has been well-defined.
2. Once an architecture has been well-defined, many useful (and interesting) views may be developed that are pertinent to fine-tuning and detailing the architecture.
3. In general, an accurate architecture cannot be reliably inferred from three views of the system/architecture or a representation of the system/architecture.

The first of these propositions suggests that the entire "views" notion from DoDAF may be questionable. There are, after all, certain logical precepts with respect to architecting. If an architecture, or a system, is X, then views of the architecture or system require that X be well defined. In other words, in order to properly derive views of an architecture, or system, that architecture, or system, needs to be defined in some appropriate detail.

The second proposition continues on and places a great deal of importance on the notion of defining, with some precision, the content of an architecture. Given that, then one may proceed with improving and providing additional detail about that architecture.

The third proposition suggests that the three views of DoDAF generally will not define, or allow one to infer, the system architecture. Let's look at an example in that regard. Suppose we wish to find out the architecture of a human being. Suppose further that we take three pictures (views) of a human being – front, side, and top. Are we then able to infer the human architecture? Or – suppose we construct the three views of operational, system, and technical. Are we then able to infer the human architecture? From the notions in this book, it is claimed that the human architecture has a lot to do with understanding and specifying the following human functions/subsystems [5]:

1. Immune
2. Circulatory
3. Nervous
4. Respiratory
5. Reproductive
6. Digestive
7. Endocrine
8. Urinary
9. Lymphatic
10. Musculoskeletal.

Thus we are brought back to the central idea and theme of functional decomposition. If one were charged with the responsibility of re-architecting the human body, the above list would certainly be the place to start.

Additional Quantitative Views

We note that the evaluation framework for this architecting procedure yields quantitative estimates of effectiveness. These are composed, in turn, of estimates for the individual evaluation criteria (e.g., maintainability, risk). This means that there is an opportunity to create quantitative views that are not otherwise available. Examples of such views are shown below [4].

1. Risk vs. Requirements
2. Cost by Function
3. Effectiveness vs. Risk
4. Effectiveness vs. Availability
5. Cost vs. Requirements
6. Risk vs. Human Factors
7. Effectiveness vs. Human Factors
8. Aesthetics vs. Risk.

Tradeoffs

Tradeoffs are detailed examinations and comparisons between the features of a system to determine which features are better, and under what conditions. They can be carried out at the architecting level, and also at the detailed (subsystem) level, in an attempt to bring the system design closer to optimal. This so-called optimality may refer to improvements in effectiveness or improvements (decreases) in cost, or both. They are well known to system designers and often require at least several weeks, and possibly months, to execute.

Some examples of tradeoffs are examined below, based upon the architecting processes in previous chapters.

The Anemometry System

We see several cases for which a more detailed tradeoff investigation might be called for. Referring to Table 3.2, we might look more carefully, and do a tradeoff study of:

- Might the radio transducer be part of the "knee-of-the-curve" solution?
- Might the custom slip rings be preferable to the stranded wire harness in the "knee-of-the-curve" architecture?
- Might the optical position sensor be a better choice than the magnetic position sensor in the "knee-of-the-curve" construct?

The generality here is that a design choice that is part of the high effectiveness solution might also be a better choice for the "knee-of-the-curve" architecture.

Architecting a House

As with the above, here are a few areas that might call for a tradeoff investigation:

- Might the den in the high effectiveness solution also be part of the "knee-of-the-curve" architecture?
- An alarm system might be added to the solution for the "knee-of-the-curve" system, and
- A sprinkler system might be part of the "knee-of-the-curve" solution.

Architecting an Automobile

- Possibly it is cost effective to add a third row with two to three seats for the high effectiveness architecture.
- Might the video assist be part of the "knee-of-the-curve" solution?
- Does a hybrid (adding electric capability) fit into the "knee-of-the-curve" and the high effectiveness solutions?

Certainly, many other tradeoff possibilities are evident by a drilling down into the synthesis tables for these three examples. These types of tradeoffs help to fine tune and improve the alternative architectures. And they have been part of classical systems engineering from its beginning.

References

1. IEEE P1471, "Recommended Practice for Architectural Descriptions," December 1999.
2. ISO/IEC/IEEE 42010, "Systems and Software Engineering – Architecture Description," November 24, 2011.
3. DoDAF Version 2.02, see dodcio.defense.gov.
4. Eisner, H., *Essentials of Project and Systems Engineering Management*, 3rd Edition, John Wiley, 2008.
5. Gray, H. and H. V. Carter, *Gray's Anatomy*, Churchill-Livingstone, Elsevier, 2008.

chapter nine

DoDAF and Other Frameworks

Department of Defense Architectural Framework (DoDAF)

The first chapter of this treatise introduced the fundamental approach set forth by DoDAF, namely, the three views of a system/architecture. These were:

1. The operational view
2. The systems view, and
3. The technical view.

This basic choice set the stage for much of what was to follow. In that sense, further advances could be characterized as "drilling down" with additional artifacts related to these three views. This was a quite reasonable approach, and in many ways a predictable one. Once a basic approach is agreed upon, we tend to look more and more deeply into the possible implications and details of that approach. The remainder of this section summarizes some of the more important aspects of this DoDAF approach, and other frameworks.

The Chief Information Officer (CIO) of the DoD had this to say about DoDAF [1]:

> DodAF is the overarching comprehensive framework and conceptual model enabling the development of architectures to facilitate the ability of DoD managers at all levels to make decisions more effectively through organized information sharing across the Department, Joint Capability Areas (JCAs), Mission, Component, and Program boundaries.

This, of course, sees DoDAF in terms of providing "information," which is largely generic, and assisting in terms of decision-making. This approach, as of Version 2 of DoDAF, emphasizes "data" in distinction to "products." The earlier product emphasis had its day and gave way to the next perspective and set of artifacts.

Part of the next perspective had to do with Models and Fit-for-Purpose Views. These are described by the CIO as follows [1]:

- *Models* are created from the subset of data for a particular purpose. Once the models are populated with data, these "views" are useful as examples for presentation purposes, and can be used as described, modified, or tailored as needed.
- *Fit-for-Purpose Views* are user-defined views of a subset of architectural data created for some specific purpose.

The DoD CIO further continues to describe the concept of an integrated architecture which is:

- An architecture consisting of multiple views or perspectives facilitating integration and promoting interoperability across capabilities and among integrated architectures.

The DoD has recognized that systems integration and interoperability are continuing issues within the DoD and that they're not likely to be "solved" any time soon.

In terms of the notion of a specific method of architecting, the CIO, in version 2.0, says that the approach "provides, but does not require, a particular methodology in architecture development." Version 2.0 also presents the notion of viewpoints, in the following dimensions:

1. All Viewpoint (AV)
2. Capability Viewpoint (CV)
3. Data and Information Viewpoint (DIV)
4. Operational Viewpoint (OV)
5. Project Viewpoint (PV)
6. Services Viewpoint (SvcV)
7. Standards Viewpoint (StdV)
8. Systems Viewpoint (SV).

All of this, as well as other artifacts, provides information for the six core processes that DoD supports. These key processes are the [1]:

1. Joint Capability Integration and Development System (JCIDS)
2. Planning, Programming, Budgeting and Execution (PPBE)
3. Acquisition System (DAS)
4. Systems Engineering (SE)
5. Operations Planning (OPLAN)
6. Capabilities Portfolio Management (CPM).

Version 2.0 of DoDAF introduced a data model known as the DoDAF Meta Model (DM2). The goal represented by this model was to provide the capability to integrate, analyze, and evaluate the systems' architectures "with more precision."

So, a bottom line with respect to DoDAF is simply that work on the architectural framework continues, and also that this work has the potential for influencing important capabilities within the DoD.

MoDAF

MoDAF is the British Ministry of Defence Architecture Framework, and it pertains to the "Enterprise" and not to systems. So it is applicable to any and all businesses. Despite this fact, it has several elements that are similar to that of the DoDAF.

The website for MoDAF provides an overview saying that the Framework is "an internationally recognized enterprise architecture framework developed by the Ministry of Defence to support defence planning and change management activities" [2].

One might say that the centerpiece of MoDAF is its M3 Model, where M3 stands for MoDAF Meta-Model. This model is based upon seven "viewpoints," namely the [2]:

1. Strategic Viewpoint (StV)
2. Operational Viewpoint (OV)
3. Service Oriented Viewpoint (SOV)
4. Systems Viewpoint (SV)
5. Acquisition Viewpoint (AcV)
6. Technical Viewpoint (TV)
7. All Viewpoint (AV).

So, if we think of any business, we see that these viewpoints represent important perspectives (views) with respect to that business.

MoDAF states that there is no specific and "official" architectural process that they endorse or require. Apparently, this is a matter of policy and the desire to leave this matter to the ingenuity of the provider. It makes sense, and this author appreciates the position that is taken. In this connection, though, MoDAF architectures "are developed as coherent, contiguous models that when viewed as a whole present a complete picture of the enterprise" [2].

TOGAF [3]

TOGAF, "The Open Group Architecture Framework," is an Open Group Standard. Its roots are in a DoD document called TAFIM, "The Architecture Framework for Information Management." The claim is that TOGAF has and maintains the de facto standard for enterprise architectures.

TOGAF 9.1 is a 700-page document that provides essential information about the TOGAF standard. This standard has three main parts: (1) the Enterprise Continuum, (2) the TOGAF Resource Base, and (3) the TOGAF Architecture Development Method (ADM). The latter is composed of key information about:

1. Architecture Vision
2. Business Architecture
3. Information Systems Architecture
4. Technology
5. Opportunities and Solutions
6. Migration Planning
7. Implementation Governance
8. Architecture Change Management.

The TOGAF provides guidance regarding how to structure and develop an enterprise architecture. It is elaborate and comprehensive. Its Forum keeps the activities of TOGAF moving forward and staying current.

The ZACHMAN Framework

John Zachman apparently developed this Framework when he was working for IBM in the 1980s. The Framework is distinctly an Enterprise model, and purports to apply to the overall business or enterprise. Thus it is supposed to explore and explain what an enterprise is all about, i.e., one can find essentially all activities of the enterprise as part of the Framework.

The core of the Zachman Framework is a matrix (two-way table) that maps the basic query questions against various activities of the enterprise [4]. The query questions are the well-known:

1. Who
2. What
3. When
4. Where
5. Why, and
6. How.

The rows of the matrix are:

1. Contextual
2. Conceptual
3. Logical
4. Physical, and
5. Detailed.

It is also claimed that a basis for the Framework lies in transforming an idea or concept into instantiations. As far as this author can tell, the Zachman approach is "alive and well" despite the apparent fact that it is not obvious as to how it "covers" the essentials of the enterprise.

Strategic View of the Enterprise

It would appear that there is considerable interest in an Enterprise Architecture, which is quite different from a Systems Architecture. For this author, an Enterprise Architecture presents areas of special importance to the enterprise. Indeed, such an architecture should depict a strategic or generalized view of the enterprise.

In that context, this author, some years ago, provided such a view [5], with the following structure:

a. Internal Perspectives
   a.1 Vision
   a.2 Culture
b. People
   b.1 High Performance Teams
   b.2 Accountability and Rewards
c. Systems
   c.1 Continuous Improvement and Reengineering
   c.2 The Learning Organization
d. Differentiation and Leverage
   d.1 Innovation
   d.2 Responsiveness
e. External Perspectives
   e.1 Customer
   e.2 Competitors

So we see several perspectives with respect to the activities of an enterprise, the sum of which might well be called the enterprise architecture.

Other Frameworks and Architectures

We note, in this section, the fact that many other frameworks and architectures have appeared in practice and the literature. For example, Sillitto has explored system architecting in his informative and leading-edge book [6]. Dickerson, along with other authors, has used architectures for research, development, and acquisition (RDA) [7]. Steven Spewak was well known for his development of Enterprise Architecture Planning (EAP) [8]. At the federal government level, there came to be a federal enterprise architecture framework (FEAF) as an initiative of the Office of Management and Budget [9]. Another relevant approach was formulated by NIST as an Enterprise Architecture Model [10]. MITRE documented its approach to dealing with architecture frameworks in its exposition of architectural frameworks, models, and views [11]. The well-known service-oriented architecture was described in a Wiley series [12], and yet another Wiley book [13], well worth examining, documented "system architectures."

In the next chapter, we explore the rather complex topic of software architecting. The reader is urged to move on, and to consider a reference [14] or two on this somewhat obscure subject.

And, of course, let's not forget where it all began – with Eberhardt Rechtin [15].

References

1. See website at dodcio.defense.gov/Library.
2. See website at www.mod.uk.
3. See TOGAF website at www.opengroup.org/togaf.
4. See Zachman website at www.zachman.com.
5. Eisner, H., *Reengineering Yourself and Your Company – From Engineer to Manager to Leader*, Artech House, 2000.
6. Sillitto, Hillary, *Architecting Systems*, College Publications (UK), 2014.
7. Dickerson, C. E., S. M. Soules, M. R. Sabins, and P. H. Charles, "Using Architectures for Research, Development and Acquisition," Department of Defense (DoD).
8. Spewak, S., *Enterprise Architecture Planning*, QED Pub. Group, 1993.
9. "Common Approach to Federal Enterprise Architecture," Office of Management and Budget (OMB), U.S. Government.
10. "NIST Enterprise Architecture Model," National Institute of Standards and Technology (NIST), U.S. Government.
11. "Architectural Frameworks, Models and Views," MITRE Corporation, see www.mitre.org, systems engineering guide.
12. Hurwitz, J., R. Bloor, M. Kaufman, and F. Halper, *Service-Oriented Architecture for Dummies*, Wiley Publishing, 2009.
13. Sage, A. and W. B. Rouse, *Handbook of Systems Engineering and Management*, John Wiley, 1999; see A. Levis, ch. 12, p. 427.
14. Taylor, R., N. Medvidovic, and E. Dashofy, *Software Architecture*, John Wiley, 2010.
15. Rechtin, E., *Systems Architecting*, Prentice-Hall, 1991.

chapter ten

Software

A natural question that follows from the previous chapters is – does the same architecting procedure apply as well to software? The answer, basically, is "yes." However, design at the subsystem level must take into account the software-unique constructs and methods. The nature of these constructs and how software fits into the overall method are explored in this chapter.

Software Engineering Architecting [1][1]

Much of the remainder of this chapter is taken from paper written by the author of this book. This paper presents a straightforward definition of a software system architecture, which can be found in the interesting treatise on agility and discipline [2] as:

> a software system architecture defines a collection of software system components, connectors and constraints; a collection of system stakeholders' need statements; and a rationale which demonstrates that the components, connectors and constraints define a system that, if implemented, would satisfy the collection of system stakeholders' need statements.

This definition clearly emphasizes components, connectors, constraints, and stakeholder needs.

Garlan and Shaw set the stage for a better understanding of software architecting as early as 1994 [3] with their overview of this important topic. At that time, it was an "emerging" field, and they articulated a number of common architectural "styles." They asserted that a number of heterogeneous styles could be combined into a single design. Examples of styles included:

- The pipe and filter style
- The data abstraction and object-oriented style
- The event-based, implicit invocation style
- The layered system style.

The software system architect has these styles at his or her disposal, and the results of the style selections, in effect, constitute an architecture. If we move to yet another source [4], we see a pointer toward software system *structure*, which

> includes the organization of a system as a composition of components; global control structures, the protocols for communication, synchronization and data access; the assignment of functionality to design elements; the composition of design elements; physical distribution; scaling and performance; and dimensions of evolution. This is the software architecture level of design.

Perhaps the most relevant paper in terms of the overall topic of this exploration is that produced by Mark Maier [5]. First, he gives us a definition of an architecture as "the fundamental organization of a system, embodied in its components, their relationship to each other and the environment, and the principles governing its design and evolution." He also cites that an architecture is the "embodiment of the set of design decisions that define essential characteristics of the system." Moving explicitly to software, he suggests that such an architecture is

> the embodiment of the earliest set of design decisions about a (software) system, and these early bindings carry weight far out of proportion to their individual gravity with respect to the system's remaining development, its service in deployment, and its maintenance life.

With respect to the importance of a coherent architecture, Maier properly asserts that "if a system has not achieved a system architecture, including its rationale, the project should not proceed to full-scale development." Further, Maier basically does not support the notion that systems and software architecting should be based upon the same or similar methods. His reasons are traceable to his observations as to the differences between system and software developments and especially their structures.

As part of his case, Maier also points to a comment by one of our most capable engineers, Frederick Brooks. In Brooks' notable treatise on software engineering [6], he points out the need for conceptual integrity in our software systems. This is typically embodied in the software architecture. A good software architect will thereby assure this integrity, i.e., it is the job of the software engineer to make sure the system has the required integrity.

Finally, with respect to the matter of software architectures, we take a brief look at the IEEE recommended practice for architectural descriptions [7]. In this "standard," a formal architectural description (AD) is introduced. Further, at that time, it was acknowledged that there was not "any reliable consensus on a precise definition of a system's architecture." However, it is still possible, and desirable, to record an architecture by its description (AD). A system's AD can also be directly related to a set of "views" of the architecture of that system. In effect, what they were saying is: we don't have consensus on what an architecture is, but we can still provide descriptions (and views) for practitioners of software engineering.

A Unified Approach

A conclusion that may be drawn may be expressed by the following:

- The current evidence is that systems architecting and software architecting appear to be on divergent paths, unless an approach is suggested that demonstrates the possibility of bringing these two notions together.

This part of this chapter suggests that systems and software architecting approaches can be "unified," at the appropriate level of what is meant by an architecture.

The overall procedure is the cost-effectiveness architecting method as described in earlier chapters. A basic notion is to accept the idea that systems (hardware, software, both, etc.) can and should be broken down into functions and subfunctions. Thus, functional decomposition becomes a critical aspect of the unified architecting procedure. The next step, of primary importance in this approach, is to construct the "synthesis matrix" which defines alternative design approaches for each and every subfunction, for (at least) three systems architectures. The design approaches explicitly cover the systems domain and also the software domain. The overall notion is depicted in Table 10.1.

After the synthesis matrix has been developed, the (three) alternative architectures are evaluated using a standard weighting and rating procedure, as presented in previous chapters. This is the "analysis" step which produces measures of the cost and effectiveness of the alternatives.

We note that in this approach we are explicitly defining and evaluating alternative architectures, with the ultimate goal of finding a cost-effective architecture (solution) for the customer(s) (stakeholders). The steps of this procedure are the same as the critical "views" in that each step is defined by an unambiguous view. Some might call this a "one-to-one" relationship such that when the step is taken, the view is automatically generated.

Above all, the functional decomposition of the system becomes the unifying element of this architecting approach. As such, it brings the system design and the software design together under a common "umbrella," the subfunction. This also assures that both system and software

Table 10.1 Synthesis Matrix for Systems and Software Architecting

Functions	Sub-functions	Architecture 1	Architecture 2	Architecture 3
1	1.1	System DA1.1-1 Software DA1.1-1	System DA1.1-2 Software DA1.1-2	System DA1.1-3 Software DA1.1-3
1	1.2	System DA1.2-1 Software DA1.2-1	System DA1.2-2 Software DA1.2-2	System DA1.2-3 Software DA1.2-3
1	1.3	System DA1.3-1 Software DA1.3-1	System DA1.3-2 Software DA1.3-2	System DA1.3-3 Software DA1.3-3
2	2.1			
2	2.2			
—	—			
—	—			
—	—			
N	N.1			
N	N.2			
N	N.3			
N	N.4	System DAN 4-1 Software DAN 4-1	System DAN 4-2 Software DAN 4-2	System DAN 4-3 Software DAN 4-3

considerations will be brought to bear for each and every subfunction, and with an understanding of the relationship between the system and the software, at that level of design.

We also note that this approach gives the software design engineer the freedom to use several different approaches (e.g., layered, object-oriented, pipe and filter) as applied to the various subfunctions. There is not (necessarily) one overall software approach for the entire system. Rather, there may be several different approaches for design at the subfunction level.

Summary, Future Actions, and Research

Suggested future actions depend upon understanding the possible implications of a unified architectural approach that works, from a practical point of view. The top-level features of that approach may be summarized as:

- A common framework that facilitates and incorporates both systems and software architecting.
- A key element of that framework is functional decomposition of the system, which typically is instantiated by both hardware and software.
- An overarching method that seeks to formulate a provably cost-effective system for the customer and stakeholders.
- Synthesis and analysis of alternative architectures, leading to the selection of a preferred architecture.

- Tested through the development of hundreds of architectures, containing both systems and software elements.
- The steps of the architecting process provide outputs that are themselves the critical views, i.e., a one-to-one correspondence between the steps and the most important views.
- The critical views are: (1) functional decomposition, (2) synthesis of alternative architectures, (3) analysis of alternative architectures, and (4) graphical representation of the cost and effectiveness of each alternative.

In terms of actions, this author suggests:

- Widespread use of the architecting method across the board within industry, academia and government
- Further research with the main focus and purpose to improve and expand the methodology
- Use of the method for real systems of both hardware and software, filling a specific need
- Systematic construction of new views, above and beyond those already considered
- Formalizing the method with respect to moving from analyses of alternative architectures to the selection of a preferred architecture.

Basically, new methods are accepted as more and more people experiment with them and find successful outcomes.

Research Areas

Back in 2006, Barry Boehm highlighted a trend that he called "the increasing integration of software and systems engineering" [8]. Given the key issue of this chapter, we would see the matter of unifying systems architecting and software architecting as a major challenge under this overall trend. More specific areas of research for bringing systems and software architecting together include the following:

1. Constructing test cases that use the unifying approach suggested here
2. Looking at interoperability issues
3. Exploring inter-relationships between the system and software design approaches at the subfunction level
4. Looking at cases for which there is significant interaction between subfunctions that are not part of the same functions
5. Defining and formalizing the processes that lead to a preferred architecture from a series of alternative architectures.

Note

1. © 2013 American Society for Engineering Education, ASEE Annual Conference Proceedings, June 23–26, Atlanta, Georgia.

References

1. Eisner, H., "Systems and Software Architecting – On Separate or Convergent Paths," ASEE Annual Conference Proceedings, June 23–26, 2013, Atlanta, GA.
2. Boehm, B. and R. Turner, *Balancing Agility and Discipline*, Addison-Wesley, 2004.
3. Garlan, D. and M. Shaw, "An Introduction to Software Architecting," CMU Software Engineering Institute, Technical Report CMU/SEI-94-TR-21, ESC-TR-94–21 (1994), SEI, Pittsburgh, PA.
4. Garlan, D., "Software Architecture," *Encyclopedia of Software Engineering*, 2nd Edition, John Marciniak (E-I-C), John Wiley, 2002, p. 1318.
5. Maier, M., "System and Software Architecture Reconciliation," *Systems Engineering*, Volume 9, Issue 2, Summer 2006.
6. Brooks, Jr., Frederick, *The Mythical Man-Month*, Addison-Wesley, 1995.
7. "Draft Recommended Practice for Ads," Software Engineering Standards Committee for Architecture Working Group, IEEE P1471/D5.2, December 1999.
8. Boehm, B., "Some Future Trends and Implications for Systems and Software Engineering Processes," *Systems Engineering*, Volume 9, Issue 1, Spring 2006.

chapter eleven

Cost Estimation

It is clear that an important part of any cost-effectiveness assessment is the accurate estimation of the costs of the alternatives. In this short chapter, we explore how one might approach this topic. In particular, we comment here on two methods, both of which have some complexities but are relatively straightforward.

Method One: Top-Level Design

With this method, we carry out a top-level design of the various alternatives and proceed with a standard costing of the various elements of the design. At that point we estimate the direct labor, hardware costs, software costs, indirect labor, overhead, G & A (General and Administrative), and profit – otherwise known as the "standard" cost model. This is also called a "standard" engineering cost approach, and usually works out well unless there are some new and advanced technologies that have to be considered. If there are, one gathers up as much information on those technologies as possible and accepts that as about as well as can be done.

Given the alternatives, it is usually the case that they have some number of common elements. This simplifies the process and very likely leads to a more accurate set of overall results. All of this depends upon how similar the alternatives are to current practices and systems. Examples are shown in Chapters 4, 5, and 6, especially the former.

The "standard" cost elements of a life cycle cost model contain the following top-level categories of cost [1, p. 268]:

- R, D, T & E (Research, Development, Test, and Evaluation)
- Procurement, and
- O & M (Operations and Maintenance).

Method Two: Cost Estimating Relationships (CERs)

This method can be rather complex as we are constructing CERs, typically based upon data that we have from existing systems. We use such data, and a statistical analysis, to formulate the cost estimates. The analysis can be linear or non-linear, with the latter usually somewhat more difficult. The former is often based upon what is called linear regression analysis. The following few lines provide the basic formulae for such an analysis [1].

If the basic linear equation is what we expect as $Y = m X + b$, we may calculate the values of "m" (slope) and "b" (y intercept) Σ as:

$$m = (n\Sigma \times y - \Sigma \times \Sigma y)/(n\Sigma x^2 - (\Sigma x)^2) \text{ and}$$

$$b = (\Sigma y \Sigma x^2 - \Sigma \times \Sigma \times y)/(n\Sigma x^2 - (\Sigma x)^2)$$

Note that the denominators are the same for both "m" and "b."

To reinforce the point, we look at a simple example. This illustration has the following six points:

X	Y	
1	2	From these points we calculate these values – $\Sigma x = 21; \Sigma y = 27; \Sigma x^2 = 91; \Sigma xy = 112$
2	3	
3	4	We then substitute these numbers and find the slope and y intercept as:
4	5	$m = 1$ and $b = 1$
5	6	and the line is $Y = mX + b = X + 1$
6	7	

There are numerous CERs that have been catalogued for immediate use, such as those listed in Table 11.1.

COCOMO I

This is the name for "Constructive Cost Model One," and it dealt with the costs of software. Going back to 1981 [2], it represented a way for engineers to estimate the costs of software, with a few simple equations [3], as below:

$$PM = C\,(KDSI)^x$$

$$TDEV = D\,(PM)^y$$

$$PROD = DSI/PM$$

$$FTES = PM/TDEV$$

Where KDSI = thousands of delivered source instructions
PM = person-months needed to complete the software
TDEV = required development time
PROD = overall productivity
FTES = full-time equivalent staff required
C, D, x, and y = empirically derived constants.

Table 11.1 Cost Estimating Variables for Various Equipment

Type of Equipment	Cost Estimating Variables
Radar Systems	Output Power Frequency Bandwidth Weight
Aircraft Engine	Thrust Bypass Ratio
Satellite Terminal	Antenna Size Output Power Frequency Receiver Sensitivity
General Radio	Power Frequency Number of Channels
Software (effort)	Source Instructions
Coaxial Cable	15.75 L where L = length of cable
SHF Earth Terminal	0.0835D + 0.157P + 0.679 where D = antenna diameter; P = power (KW)

For a software development situation characterized by a small team, considerable experience and a stable environment (Boehm's organic mode [2]), the above equations can be re-written as:

$$PM = 2.4\ (KDSI)^{1.05}$$

$$TDEV = 2.5\ (PM)^{0.38}$$

So, we "input" the estimated KDSI and are able to immediately develop estimated values for person-months (PM) and development time (TDEV). This can be illustrated by taking the value of KDSI to be 80. We then have:

$$PM = 2.4\ (80)^{1.05} = 2.4\ (99.6) = 239 \text{ person-months, and}$$

$$TDEV = 2.5\ (239)^{0.38} = 2.5\ (8.01) = 20.03 \text{ months}$$

PROD = 80,000/239 = 334.7 DSI/PM, and for 20 days per month,
PROD = 16.7 DSI/day

The latter number can be examined to see if it is reasonable from one's own experience. Also,

FTES = PM/TDEV = 239/20.03 = 11.93 or about 12 people, equivalent.

COCOMO II [4]

This is an extension of COCOMO I, again led by one of our most extraordinary software engineers, Barry Boehm. The basic formula for calculating person-months is the same as for COCOMO I, namely:

$$\text{Effort Required} = \text{PM} = \text{A (size)}^{\text{B}}$$

where A is a function of EMs, or effort multipliers, and B is related to a set of scale factors. There are seven or 17 EMs that relate to A, and several scale factors that are spelled out in the basic COCOMO II text and represented by such "variables" as risk resolution, team cohesion, and others.

Changing Costs

In the context of comparing costs in a cost-effectiveness evaluation, we are most interested in relative costs, rather than absolute costs. In this real world of ours, we often see quite large increases in cost estimates as the years pass and as the original assumptions for our systems change. This is to be expected, together with the headlines that reveal the timing and magnitudes of these changes.

In recent times, we can look back upon the history of the Joint Strike Fighter and find major increases in cost that have been experienced and reported in the various journals. There are many reasons for these increases, a large number of them entirely understandable as well as justifiable. One needs to keep in mind that choosing the best system, or architecture for that system, is quite different from assuring accurate estimates of costs for these systems. Both are also different from being able to control costs and the variables that might lead to cost increases over time.

Cost Information Sources

The architecting team usually has considerable technical information available, but often lacks cost information. However, several government agencies have tried to "fill the gap" by providing handbooks of cost information of various types, such as design, operating costs, and maintenance costs. Two such agencies are NASA [5] and the GAO [6]. It is usually a good idea for some members of the system architecting team to retrieve this type of information rather than attempt to construct it from scratch.

During the days of Secretary Robert McNamara in the DoD many person-years went into the task of finding and making available cost data for analyzing military systems and their cost effectiveness. The Rand Corporation, it can be remembered, had steady contracts to support the DoD's work in this area. It is also remembered that Alain Einthoven, an economist, had a key role to play to try to assure that all cost information was treated appropriately.

References

1. Eisner, H., *Computer-Aided Systems Engineering*, Prentice-Hall, 1988.
2. Boehm, B., *Software Engineering Economics*, Prentice-Hall, 1981.
3. Eisner, H., *Essentials of Project and Systems Engineering Management*, 3rd Edition, John Wiley, 2008, p. 320.
4. Boehm, B., C. Abts, A. Winsor Brown, S. Chulani, B. K. Clark, E. Horowitz, R. Madachy, D. Reifer, and B. Steece, *Software Cost Estimation with COCOMO II*, Prentice-Hall, 2000.
5. NASA Cost Estimating Handbook, v. 4.0.
6. GAO Cost Estimating and Assessment.

chapter twelve

Summary

This last chapter summarizes many of the key points that are made in this treatise.

I Basic Elements of the System Architecting Process

Although much has been written about how to architect a system, there remains considerable lack of clarity in the literature about the specifics of such a process. This monograph is quite specific about how to develop a system architecture. Four elements are set forth:

1. Functional Decomposition
2. Design Approaches to Instantiate All Functions and Subfunctions (Synthesis)
3. Evaluation of Alternatives (Analysis)
4. Selection of Preferred Alternative (Cost-Effectiveness Assessment).

We further note that some important aspects of systems engineering appear to not be included in the above list, such as "requirements analysis." This is deliberate and part of the philosophy of the process (see "Rechtin" citation below).

II The Importance of a Cost-Effectiveness Evaluation

We note that the overall systems architecting procedure delineated in this book is a "cost-effectiveness" evaluation. This means that the procedure has a strong body of literature to rely and depend upon. It also is intuitively appealing in terms of the various features of such an evaluation (such as a choice from among a set of alternatives, how to develop and use measures of effectiveness, and others). It also means that we are looking for a "cost-effective" solution to the system design problem.

III Functional Decomposition

We further note that the first step in the process of architecting is "functional decomposition." This is a crucial first step as it sets the stage for both synthesis and analysis by being precise about the functions and subfunctions that are part of the system. The listed functions represent the minimum set that are part of the system. All alternatives must have these functions, but some of the alternatives may have increased functionality that increase performance and require additional funds.

There is some question about how many levels the functional decomposition should contain. If the name of the system is at level "zero," then two additional levels are recommended for purposes of system architecting. This suggestion may not apply when one is considering a "system of systems."

IV The Synthesis Step Reveals and Defines the Architecture

The second step, synthesis, is indeed the definition of the architecture. It shows, for each system function and subfunction, the design approach that is selected by the system architects. This step makes it quite clear as to what the design choices are. Further, these choices are explicit for at least three system approaches: (1) the low-cost, low-effectiveness system, (2) the high-cost, high-effectiveness system, and (3) the knee-of-the-curve, best value system. Often, we are in search of the latter system as the preferred system. The important point here is that we are not leaping to the conclusion that we go directly to the selected alternative; rather we define and ultimately evaluate a set of alternatives.

V Rechtin's "Kiss" Notion

Eberhardt Rechtin wrote a seminal book on the subject of "system architecting" [1]. In that treatise he articulated many "heuristics," including that of the "KISS" (keep it simple, stupid) approach. This was based upon his many years and experiences as a master engineer and builder of systems. This principle is accepted as the preferred approach to systems architecting. Yet another observation he made was to keep in mind Occam's Razor, which said that "the simplest solution is usually the correct one."

Another aspect of "keeping it simple" is that the four architecting steps can be displayed on one, and only one, page each. This too is deliberate and assists in lending clarity to the process, especially in a group setting.

VI Problems with "Views"

The approach to systems architecture selected by the Department of Defense, otherwise known as DoD Architectural Framework (DoDAF) [2], was based upon three views of an architecture, namely:

1. The operational view
2. The systems view, and
3. The technical view.

For this author, this approach has several problems. First, one is not able, in general, to develop a view of an architecture without first having an architecture in hand. Second, one cannot infer an architecture, necessarily, from these three views. As an example, if one wishes to develop an architecture for the human body, how do you get there from the above three views, or indeed from the "standard" three views of any system? At the same time, the architecting procedure suggested here provides a framework for developing new and relevant views that assist in the overall evaluation process. Some of these views are defined in Chapter 8.

VII Centrality of Defining Alternatives

As noted above, the overall approach to system architecting in this book is based upon the notion that it is important to define alternatives as part of the process. This is critical and in concert with approaches generally taken by the DoD [3] in the design of systems.

VIII Lateral vs. Drilling Down Approach

The above approach, in terms of defining alternatives, is considered "lateral" in distinction to "vertical" or "drilling down." In the lateral procedure, we specifically define and look at alternatives. In the drill down approach, we leap to one alternative and develop lots of data about that alternative. This is not a recommended procedure for design, and often leads to huge amounts of irrelevant information.

IX A Framework for Tradeoffs and Sensitivities

It is well known that an important part of any design process is to carry out a variety of "tradeoff" analyses and investigations in order to seek the best solutions at the system and well as subsystem level. The relatively simple structure, especially of the "synthesis," facilitates these tradeoffs. Areas for tradeoffs can be seen directly by examining the design choices for the various functions and subfunctions. We look down the columns to explore compatibilities and interoperability matters. Then we look across the alternative columns to discover tradeoff areas. In all cases, we are looking to define fruitful tradeoffs, and set the stage for a deeper investigation of how system features and parameters inter-relate.

This framework for architecting also facilitates what many call a "sensitivity analysis" in addition to the tradeoffs suggested above. For example, modifications in the ratings as well as the weights for the evaluation criteria reveal how sensitive the overall results are to these changes.

X Software System Architecting

An important aspect of system architecting is that of architecting software systems. For the approach in this monograph, it is suggested that the "functional decomposition" step is the integrating element that brings together both hardware and software system architecting. This is based upon many years of exploring and defining how this may be executed. The way that this may be achieved is cited in the text and the graphic in the chapter on "software." The specifics of the complex topic of software architecting are presented in several texts available to the software designer and engineer.

X1 A Time Constrained Preliminary Design Process

The overall context for this exploration of system architecting has two main aspects, namely:

1. The limited time frame
2. More-or-less equating system architecting with preliminary design.

The limited time frame, of the order of a month or two, suggests a limited set of analyses. It also means that the procedure is especially suited to the proposal-writing process. We are constantly in search of ways and means of improving our proposals.

Beyond this time constraint, we are equating system architecting with preliminary system design. There is really no need to try to differentiate between these two notions.

X11 System Architecting as a Group Process

System architecting is so important in terms of building new systems that it must be carried out by a group. The recommended size of the group is eight to 12 people (depending upon the size of the system) holding highly interactive design sessions on a regular and rapid basis. Each person that is a member of the team must be senior and highly experienced with the type of system under consideration. The members must also have proven records as "team players," i.e., they must understand how a high-powered team is supposed to operate and have signed on to that proposition.

Of particular importance is the selection of the team leader. Such a person must have extraordinary design experience and know how to "manage" a group of very senior personnel. Such a person will have gained the respect of several teams as part of his or her background. Such a person will have demonstrated that he or she is a true leader in both technical and management areas. There is quite a lot of literature on this matter of group operation and leadership that can be helpful in this domain.

References

1. Rechtin, Eberhardt, *Systems Architecting*, Prentice-Hall, 1991.
2. DoDAF Version 2.02, see dodcio.defense.gov.
3. "Analysis of Alternatives (AoA) Handbook," see www.prim.osd.mil.

appendix a

Group Architecting

There has been some mention of group processes in this book. Here we focus upon various ways that a group might carry out the suggested architecting procedure, and some ways that are likely to be counter-productive. The procedure itself can, of course, be executed by a single person, an individual. For improved results, it's best to explore the process in the hands of a group, and, especially, a group leader. Having mentioned a group leader, we reiterate that a crucial part of the process is to assure a most capable group leader. In the view of this author, "it's all about effective leadership" to move a group process from potentially dysfunctional to high performance. That's the ultimate goal – to have an outstanding, high performing group process that deals with this most important aspect of building systems – developing the architecture.

It all starts with designating a most effective team leader to be in charge of the group architecting process. The general characteristics of a strong leader are discussed in some detail in a previous book [1] by the author. Assuming that this is achieved, we ask ourselves what it is that this leader starts with. And the answer seems to be to set up the first group architecting session, with an agenda that:

1. Calls for a complete commitment to the process (i.e., forms the architecting team)
2. Explains the overall four-step architecting process (see Chapter 1)
3. Emphasizes the "synthesis" step
4. Requires that all participants have a specific assignment dealing with that synthesis step
5. Insists upon appropriate participant behavior in terms of following the leader's ground rules, and accepting the agenda.

It is quite important to recognize that a critical part of "synthesis" is the proper instantiation of each and every function, with a design approach. So – if there are six functions, we need a "synthesizer" for each. If there are 12 functions, we need a synthesizer for each, as well. In the latter case, therefore, we need about 17 people to form the initial architecting team – 12 for the 12 functions, three more of your best systems engineers, and another two with specific skills in "cost" analysis. That's a large team, and it needs to be disciplined in order to succeed. A strong leader, as suggested above, has the job of keeping this team working effectively and with appropriate focus.

The pace of progress and the dates for follow-up meetings, are entirely in the hands of the leader, as is the overall schedule for the project. The high priority for this activity should ensure that all team members are present for follow-up meetings, and also that they have made the required "technical" progress (i.e., have defined, in some detail, the design approach for each function). Full-scale review and discussion of these results are part of the next meeting's agenda. It's obvious that this can take a lot of time. All members of the team need to clear the deck and provide their best efforts in this endeavor. This is not a "pushover" assignment – it's one that is challenging right from the beginning. The system architecting sets the stage for much of the engineering design that follows – both in concept and in detail. And as we approach the first team meeting, we briefly cite what we expect from the leader as we navigate these meetings:

1. Respectful listening to each and every member of the team
2. Summary as to "where we are" at appropriate points in the meeting
3. Reminders as to the progress that we need to make in order to meet the goal of the project and the overall goals of the enterprise
4. Keeping certain members of the team in check when it's appropriate to do so
5. Showing "group" displays that can be seen at the same time by all participants
6. Making sure that "architecture costs" are not neglected, which may happen with a purely technical group
7. Keeping administrative costs relatively low.

The First Team Meeting

The first overall meeting of the team should be about one to two weeks after establishing the team. The primary focus of that meeting, as suggested above, is a preliminary overall architecture, formed as a one-page synthesis chart. A required input is needed for that meeting from each functional area, in terms of a design approach. Also, on the other axis, the same is true for a "cost" for each architecture. All of this is very preliminary since the participants have had only one to two weeks to analyze the problem and prepare their material. At the meeting, scheduled for two to three hours (depending upon the number of functional areas), there is a free-wheeling discussion of the results, their rationale, supporting data, criticism, alternatives, new data from new sources, the cost dimension, the need (or not) for some sensitivity analysis, and whatever the team leader accepts as relevant new inputs or data. The overall goal of the results of that meeting is to develop a very preliminary "best" set of architectures along with a series of steps upon which to focus for the next and follow-up meetings. Although it's important to try to plan for these meetings, we keep in mind that a strong and effective leader has it all in his or her hands; they are able to shift direction instantly when they see a need for it. Leaders know how to do this.

We close here with a few comments about effective group interaction and problem-solving, as well as behavior to avoid.

Group Problem-Solving

Here are a half-dozen pointers pertaining to group interaction and problem-solving:

1. Assure that all participants are attentive and listening to all group interactions.
2. In view of the above, don't allow separate discussions that tend to subvert the group process.
3. Insist that all participants "do their homework."
4. Insist that all participants be respectful of one another.
5. Go around the room and explicitly ask for responses and inputs on the matter at hand.
6. Start and end each meeting with very brief summaries of the past and anticipated meetings, with an emphasis on goals approached and achieved.

Behavior to Avoid

Here are some six suggestions for behavior to keep away from:

1. Do not allow any one participant to dominate the discussion, even if the points made are useful and coherent.
2. Don't allow side discussions that are likely to draw away from strong group behavior and participation.
3. Quickly stop team "busters" that attempt to subvert the leader's direction and control.
4. Don't allow the discussion to get "off course" in terms of the meeting's agenda.
5. Keep away from combining group participation and a grand "lunch."
6. Respect the timetables that have been set for the topics and the overall meetings.

GroupThink

We especially need to avoid "groupthink," which is "the tendency in groups for a convergence of ideas and a sanctioning of aberrant ones" [2]. Experienced group leaders are well tuned to this type of group behavior and know how to keep it from happening.

References

1. Eisner, H., *Reengineering Yourself and Your Company*, Artech House, 2000, ch. 6.
2. Sage, A. and W. Rouse, *Handbook of Systems Engineering and Management*, John Wiley, 1999, p. 663.

appendix b

Functional Decomposition

Functional decomposition plays such a central role in architecting systems that it deserves a special place and discussion; hence this Appendix. Examples are provided here, in the four fields of (a) information technology (IT), (b) communications, (c) space systems, and (d) transportation.

Information Technology (IT)

There are many types of IT systems and so the functional decomposition may be considered somewhat generic, as below:

1. Input
2. Output
3. Processing
4. Storage
5. Security
6. Database management
7. Power supply
8. Operating System (OS)
9. Applications.

The first two of the above are directly recognizable, with subfunctions like mouse, keyboard, screen, and printer. The "critical" part of the adventure, as far as most users are concerned, is the "processing" function as they are looking for speed. Storage is relatively simple to come by these days, and "security" is a necessity with the intrusions of various types that we might anticipate. We include database management as a generic function and we may suppose that this function will ultimately be instantiated with a commercial DBMS such as Oracle or Access. There is very little point in building one's own and new DBMS when excellent commercial products are available off-the-shelf. Of course, all systems of this type need a power supply and incorporate a number (usually several) of applications.

There are nine functions cited above, but the work that will ultimately be done on such a system is not equally divided between these functions. The so-called "degree of difficulty" depends strongly upon the requirements for each function and also the extent to which a commercial product is available to satisfy the requirements. And we also need to take into account the number and degree of difficulty of the individual applications.

Communications

Communication systems are a special type of IT, with specialized functions, as enumerated in Chapter 3, and repeated here:

1. Multiplex/Demux
2. Modulation/Demod
3. Switching and Routing
4. Encryption/Decryption
5. Formatting/Signal Conversion
6. Control and Monitoring
7. Recording and Playback
8. Satellite/Terrestrial Communications.

Of course, there are other forms and types of communications system, with the above being only one illustrative set. To cite just one, in certain situations, we see troposcatter communications systems as well as others that distinguish between analog and digital, even though such a distinction may not be appropriate at this level of design and specification. Yet another specialized communications system was mentioned in Chapter 1, namely, the Mallard system. That was distinctly a "battlefield" communications system, with subfunctions attendant to that location and environment.

Space Systems

If we are dealing with a satellite that we are putting into orbit, we typically see the following top-level functions:

1. Power Supply
2. Stabilization and Control
3. Thermal Control
4. Communications and Data Handling
5. Telemetry (may also be part of (4) above)
6. Payload Functions (several, such as cameras and a variety of sensors).

This is, of course, just the space segment, and does not include the ground segment. Depending upon the nature and purpose of the satellite, the ground segment can be massive and extensive. Given that the satellites that are part of the 1980s SDI program were plentiful, one can only imagine what the ground stations looked like.

Another example of the functional decomposition of a "space" system pertains to the Earth Observing Data and Information System (EOSDIS) [1]. In this case, there are three Segments, namely:

1. The Flight Operations Segment
2. The Science Data Processing Segment, and
3. The Communications and System Management Segment.

For the sake of completeness, we include here the functions for each of these segments:

Flight Operations Segment

1.1 Mission Control
1.2 Mission Planning and Scheduling
1.3 Instrument Command Support
1.4 Mission Operations

Science Data Processing Segment

2.1 Data Processing
2.2 Data Archiving
2.3 Data Distribution
2.4 Data Information Management
2.5 User Support for Data Information
2.6 User Support for Data Requests
2.7 User Support for Data Acquisition and Processing Requests

Communications and System Management Segment

3.1 Distribution of EOS Data and Information to EOSDIS Nodes
3.2 Distribution of Data Among Active Archives
3.3 Interface With External Networks
3.4 Network/Communications Management and Services
3.5 System Configuration Management
3.6 System/Site/Elements Processing Assignment and Scheduling
3.7 System Performance, Fault, and Security Management
3.8 Accounting and Billing

We note how detailed the functions are at the subfunction level. However, they are usually well conceived, far-ranging, and thoroughly necessary.

Transportation Systems

Functional decomposition in this domain can depend strongly upon what type of transportation system we are considering. However, we will attempt here to be as generic as possible. Here are some functions to be explored and cited in this type of system:

1. Overall Motive Power
2. Distributed Motive Power and Network
3. Freight Compartment Features
4. PAX Compartment Features
5. Compartment Coupling, Movement, and Control
6. Intermodal Interfaces and Control
7. Safety
8. Security.

References

1. Phase C/D Requirements Specification for the EOSDIS, (1990), Greenbelt, MD, Goddard Space Flight Center.

appendix c

Special Topics

This appendix briefly discusses a variety of topics that are related to the matter of architecting and implementing large-scale systems. They further illuminate the overall architecting procedure and also provide some context for the use of such a procedure.

Rechtin's Heuristics

We return to the wise words of Eberhardt Rechtin, an original architect of systems, and take a brief look at some of his suggestions in the form of his "heuristics list" [1].

Keep It Simple, Stupid

There is considerable emphasis on the "KISS" principle since it has proven to be wise advice. The more complicated a design, the more likely that something will go wrong. The more complex a process, the more likely we will see some form of human error. This procedure, in its best form, involves a group process. These can be problematic, as we know. Even though a group may reach consensus, it does not mean that this consensus is the "right" answer. On this same point, Rechtin lists "Occam's Razor" as one of his heuristics. His statement is "The simplest solution is usually the correct one."

Serious Mistakes

Rechtin's heuristic is that in architecting a new (software) program, all the serious mistakes are made on the first day. We can interpret this suggestion in terms of getting the overall architecture right since architecting is more-or-less the pivotal point in preliminary design.

Small Number of Documents

Rechtin points to a small number of documents as critical on a project. The suggested architecting procedure can be described, literally, in four sheets of paper.

Choice Between Architectures

This heuristic, first and foremost, declares that it is necessary to look at several architectures and then pick the one that has a set of drawbacks that the client can handle best. Another related heuristic has to do with maintaining open options as long as possible in the design of complex systems.

Levels of Decomposition

The suggested architecting procedure depends heavily on functional decomposition as the first step. So the question arises: how many levels of functional decomposition are appropriate for this architecting procedure? If we call level zero the name of the system, then the recommended number of levels, for most (not all) systems, is two. This means that we are architecting at the subfunction level, i.e., each function is broken down into one or more subfunctions. For quite large systems (e.g., systems-of-systems), a further breakdown is necessary. An example of a quite large system is the National Aviation System.

A quick anecdote about this matter of levels of decomposition. This author was talking to an executive in a "systems" company and that executive offered some reasons why an important project had major troubles. His answer was that his company approach involved too many levels of decomposition, which led his design engineers down too many diverse paths. "They were deep down in the details before they had properly architected the system," he said. That led to too much unproductive spending, which put the project in jeopardy.

Drilling Down

Our reliance on "analysis" as a tool for design over the years has led us to overemphasize the process of "drilling down" in search of more detail. We note that there is not a lot of drilling down in the recommended procedure. This is deliberate, and part of the notion of "keeping it simple" at the appropriate level of design. If anything, the recommended procedure relies more on getting the functional decomposition correct and complete, which is more like a process of lateral thinking, as suggested by de Bono [2]. Try to confirm and verify that the functional decomposition is the best representation it can be, so as to avoid unnecessary searches for more detail in the downward (sub-subsystem) direction.

Dependencies and Interactions

Since the functional decomposition separates the functions from one another, the initial cut does not focus upon the possible dependencies and interactions *between* functional and subfunctions. These should not be neglected. Keep a running list of such dependencies and account for them as the "synthesis" part of the process is undertaken. This should be a simple matter of starting and keeping a list of each dependency and a note on the nature of that dependency. This could have a major influence on the design alternatives that are considered for each subfunction. This is also related to the ultimate matter of interoperability between systems and subsystems.

Costing

Not much attention is paid to the issue of cost estimation as an important part of the suggested architecting procedure. This is deliberate in the sense that we already know a lot about how to estimate costs for all types of systems. In general, we are not creating new ways of costing systems; we are mainly using existing procedures (Chapter 11). Such procedures often lean upon cost estimating relationships (CERs) [3]. These are shorthand methods that lead to cost estimates based upon existing systems and technologies. So, if we are pushing the state-of-the-art in these areas, there may be a weakness in our cost estimating capabilities and knowledge. If this is the case, multiple estimates may be called for, trying to bracket the range of costs for the various alternative architectures under consideration. Special quick studies of specialized areas may also be called for (e.g., advanced sensors, new concepts).

Requirements and Function Creep

For the architecting procedure suggested here, the requirements for the three architectures are the same. That is, all three architectures are to satisfy a given, and the same, set of requirements. In the real world, if one has been party to a system development for more than two years, as an example, requirements tend to creep. The program manager for the developer needs to be aware of such a tendency, and either resist this type of change or accept it and account for it.

In the world of preliminary design, or system architecting, as described here, there might be a tendency to have "function creep." That is, in moving from system A to system B to system C, there might be a tendency to add functionality (i.e., System B carries out more functions than does system A and system C has more functions than system B). This may be illustrated by looking more deeply at the architecting example in Chapter 5 (the house example). The high effectiveness house has alarms, internet service, a sprinkler system, and a library/video room. Thus, system C provides additional functionality for which an increase in price is expected. If one wishes to be more rigorous in one's approach, the functionality should remain the same for all three alternatives. In that way, the comparison between the three alternatives would also be more rigorous (i.e., using a so-called "apples" to "apples" comparison).

Proposals and Acquisition Ground Rules

Formulating an architecture, in a short period of time, is a critical part of many proposals to develop systems. Week after week, proposals are submitted to the federal government all of which have first-order architectures for systems in response to an RFP (request for proposal). Therefore, being able to set forth one or more architectures in these proposals is an extremely important aspect of competing for business in that (as well as other) marketplaces. What follows are some considerations with respect to developing proposals for the U.S. government.

The proposed architecture for the system is likely to be a critical section of the proposal. It may well be the most important "discriminator," i.e., the part of your proposal that positively discriminates your proposal from those of your competitors. Consider a process with the following elements:

1. Designation of your best systems engineers as your architecting group (team).
2. The above team will have at least six members, and possibly more than a dozen.
3. The above team will have a leader who has demonstrated an ability to lead a group and establish positive team behavior.
4. Each member of the active team has one or more people who serve as back-ups.
5. Lay out a master schedule for meetings and milestones for team meetings; the schedule must be consistent with the overall proposal development and completion schedule.
6. Team meetings can last all day if necessary, in order to develop the required architectures.
7. At least one member of the team will be designated as the lead writer of text in the proposal document itself; this person will be a proven excellent writer (no novices).
8. Management will spend some time sitting in on the meetings of the teams; the company cannot afford to have a bad architecting explanation and process.
9. The proposed architectures shall be clearly cost effective, reflecting minimum risk for the customer and all stakeholders.
10. Consider the classical use of a "Red Team" and allow time for this review to be accomplished.
11. The ultimate architecting result will be excellent and complete for both the team leader and management.

The proposal shall be clearly responsive to the acquisition ground rules set forth by the acquisition agent. The RFP evaluation criteria must be considered, in detail. The sections of the proposal must reflect these criteria in both form and substance. If the RFP allows it, consider submitting a proposal that contains *both* a low-cost approach as well as a "best value" approach in terms of the system architecture. Consider the list of "Ten Key Points" cited by this author in terms of developing and writing a winning proposal [3, p. 179].

Group Processes

The above proposal activities involve a group process in order to carry out the system architecting. As indicated, a strong and proven team leader is essential to a productive and successful group process. However, despite a strong leader, there may be forces at work that could threaten the result. One such force is the possible presence of a team "buster." Such a person basically has a hidden agenda that amounts to trying to sabotage the team and thereby threaten the results. The designated team leader needs to know how to deal with such a person. For a true team buster, advice from this author is to remove such a person from the team as soon as possible.

There are many other factors that relate to how a group might behave. These factors will not be examined in the main body of this treatise. The literature contains many excellent papers and books on this subject. Suffice it to say that since the system architectures are a key product and are developed in a group context, it is essential that productive group behavior is assured to the maximum extent possible. Here is a short list of items to be considered as set forth by the group leader.

1. Make sure that everyone knows everyone else on the team; use several team gatherings if necessary to get acquainted.
2. Establish and distribute a list of dos and don'ts in terms of group behavior as part of the team.
3. Such a list should include (but not be limited to) courteous interpersonal behavior, with no interruptions and in definitive listening.
4. Everyone agrees to abide by the final decisions of the team leader.

Stakeholders

It must be recognized that the ultimate test of whether or not a particular architecture is viable is acceptance by the community of stakeholders. This community tends to have many and varied constituents, which may be changing with time and what the project status happens to be. During the proposal phase, the assumption is that the key stakeholders are the members of the formal proposal evaluation team. This team decides which proposal is best and acts as a surrogate for yet another team, the ultimate system customer. The proposal evaluation team may believe that you have determined the best system architectures. But after the contract is let, the ultimate customer is to be listened to and changes may have to be accepted as new stakeholders have come upon the scene. These include changes in the architectures themselves. If you agree with such changes, all is well. If you don't agree, then one needs to be persuasive regarding the costs and effectiveness of the architectures and the benefits of your preferred selection. The bottom line is that there may be many stakeholders that have their opinions about the system architecture (as well as other project factors) and their views need to be addressed to the maximum extent possible and practical. Your internal team may believe that you have the best solution(s), but external folks may have different points of view.

What this Method (of Architecting) Is, and Is Not

To conclude this appendix, this is *not* a book about:

1. The deficiencies of DoDAF (or MoDAF...)
2. The wisdom of the fundamental "views" approach
3. Extending and expanding the three basic and early views of DoDAF
4. How to develop an architecture from the above three views
5. Why and how enterprise architecting is not the same as systems architecting
6. Drilling down to obtain more and more detail
7. Why digging more deeply is the preferred approach to better architectures
8. Fighting lack of knowledge with additional complexity
9. Additional data structures and needs for purposes of architecting
10. The benefits of group architecting.

but *is* a book about:

1. The very early activity known as preliminary design, or first-order systems architecting
2. How to do rapid system architecting, useful especially during the proposal phase of a project or program
3. Using the KISS principle to one's serious practical advantage
4. A new and compact procedure for front-end system architecting
5. Defining alternative architectures, all on one sheet of paper
6. Using lateral thinking, instead of digging deeper, to solve an important problem
7. Using cost-effectiveness considerations to develop and evaluate alternative architectures

8. Systems architecting in a group/team setting
9. Developing "views" that are the same as the outputs of the steps of the architecting process
10. A definitive four-step process of system architecting that has been demonstrated in both industry and academia, over more than two decades.

References

1. Rechtin, E., *Systems Architecting*, Prentice-Hall, 1991.
2. deBono, E., *The Use of Lateral Thinking*, Pelican Books, 1971.
3. Eisner, H., *Essentials of Project and Systems Engineering Management*, 3rd Edition, John Wiley, 2008.

Index

Printed in the United States
by Baker & Taylor Publisher Services